A. M. Legendre, Lawrence S. Benson

Geometry

The Elements of Euclid and Legendre Simplified and Arranged to Exclude from

Geomtrical Reasoning the Reductio ad Absurdum

A. M. Legendre, Lawrence S. Benson

Geometry
The Elements of Euclid and Legendre Simplified and Arranged to Exclude from Geomtrical Reasoning the Reductio ad Absurdum

ISBN/EAN: 9783337154653

Printed in Europe, USA, Canada, Australia, Japan

Cover: Foto ©berggeist007 / pixelio.de

More available books at **www.hansebooks.com**

GEOMETRY:

THE ELEMENTS OF EUCLID AND LEGENDRE

SIMPLIFIED AND ARRANGED

TO EXCLUDE FROM GEOMETRICAL REASONING

The Reductio Ad Absurdum;

WITH THE

ELEMENTS OF PLANE AND SPHERICAL TRIGONOMETRY,

AND

EXERCISES IN ELEMENTARY GEOMETRY AND TRIGONOMETRY.

ADAPTED FOR SCHOOLS AND COLLEGES.

BY

LAWRENCE S. BENSON,

Author of "The Truth of the Bible Upheld,"—London, 1864; "Geometrical Disquisitions,"—London, 1864; "Scientific Disquisitions Concerning the Circle and Ellipse," 1862.
Member of the New York Association for the Advancement of Science and Art; Hon. Mem. Phi Kappa Society, University of Georgia; Brothers' Society, Yale College, etc., etc.

NEW YORK:
PUBLISHED FOR THE AUTHOR BY DAVIES & KENT,
No. 183 WILLIAM STREET.
1867.

TO

PROFESSOR GERARDUS BEEKMAN DOCHARTY, LL.D.,

COLLEGE OF THE CITY OF NEW YORK.

———

Sir—In permitting me to inscribe to you this Treatise of Elementary Geometry, you do me great honor. Your experience and success as a Teacher and an Author will readily enable you to give a full scrutiny to the design and compass of this volume. Much originality can not be expected in a subject which has been, for more than two thousand years, enriched by a great number of eminent men; but in these days of practicability, a modification of this science may be attempted, as you have yourself thought proper to do, with a view of utilizing the important principles of Geometry, and presenting them in such a manner that though "no royal road to Geometry" can be found, the path to a knowledge of it may be rendered so clear that the impediments will be in the learner himself. And to remove much difficulty in acquiring an easy acquaintance with its numerous theorems and problems, I have thought proper to exclude the inelegant *Reductio ad absurdum* from the methods of geometrical reasoning which you have expressed—" *a consummation most devoutly to be wished,*" and which accomplishment, resulting from my labors, I now present for the benefit and use of those whose education is in the future.

I have the honor to be,

Very respectfully, yours,

LAWRENCE S. BENSON.

New York, *April 16th,* 1867.

TESTIMONIALS.

THE COLLEGE OF THE CITY OF NEW YORK, }
Cor. Lexington Avenue and 23d Street. }

NEW YORK, *January 3d*, 1867.

I have had several interviews with Mr. Lawrence S. Benson on scientific subjects, and from his conversation, together with the Essays which he has published, I esteem him an excellent scholar and fine mathematician. He has a desire to establish the Elements of Euclid *in all cases*, independently of the demonstration known as the *Reductio ad absurdum*, " a consummation devoutly to be wished."

Whatever aid or advice you can render him in the furtherance of this object will tend to the advancement of true science.

Yours truly,
G. B. DOCHARTY.

ROOMS OF THE NEW YORK ASSOCIATION FOR THE ADVANCEMENT }
OF SCIENCE AND ART, *February 28th*, 1867. }

Extract from the transactions of the Association for the Advancement of Science and the Arts:

"At a meeting of the New York Association held February 25, 1867, a paper on a new method of demonstrating the propositions of Geometry, denominated the *Direct Method*, in place of the one now in use, and called the *Indirect Method*, was read by Lawrence S. Benson, Esq., which method the writer proposes to introduce into Schools and Academies.

"After the reading of the paper, and the discussion of its merits, the subject was referred to Professor Fox, Principal of the Department of Free Schools of Cooper Union, and to Professor Cleveland Abbe, for examination and report. It was also moved and carried that the Report when received be referred to the Section on Physical Science for final disposition.

"The Section, after reading the Report of Professor Fox, the letter of Professor Abbe, and the opinion of Professor Docharty, who had been invited to examine the work, feel justified in commending this work as worthy of patronage. Professor Fox in his Report says: 'The design of arranging the Definitions, Axioms, and Propositions of Geometry, so as to use only the Direct Method of demonstration, is a good one, and when arranged in the form of a neat elementary text-book, will doubtless do much good, as the Direct Method is much more easily understood than the Indirect Method, by beginners.'

" L. D. GALE, M.D.,
Gen. Sec. of the New York Association for the
Advancement of Science and the Arts."

PREFACE.

By way of preface, I will state what I have done in this edition, and explain the reason why I have done so. I have used such propositions only which are required to substantiate the principal theorems and problems by which the principles of Geometry have practical applications in Trigonometry, Surveying, Mechanics, Engineering, Navigation, and Astronomy. I have generalized the various propositions found in the school editions of Geometry, and where particular cases arise under such general propositions, I have given the demonstrations for them. I have arranged the propositions to give the Direct Method of demonstration in place of the *Reductio ad absurdum* or Indirect Method.

My reasons for the foregoing changes are obvious to the experienced mind; considering the extent and variety of modern education, the time devoted for the pupils to acquire rudimental knowledge becomes encroached upon in order to make them acquainted with its numerous modifications; and for the pupils to obtain such knowledge of the rudiments as will enable them to see the practical applications throughout all their extent and variety, which are very great in these days of advancement and civilization, the rudiments which were taught centuries ago must be so abbreviated as to contain the *essentials* only. When materials for instructing the mind were scant, there were no opportunities to make close selection of them; but now, when those materials are plentiful, a judicious selection of the best becomes imperatively necessary. And when geometrical principles have become extended by the Algebraic Analysis, and have been made practicable by Trigonometry, Surveying, Mechanics, Engineering, Navigation, and Astronomy, the mere mental exercises, which were regarded so beneficent by the ancients, are unsuited for this practical age, which is continually bent on progress, while the intellect is sufficiently exercised by utilizing modern acquisitions; for this reason, I have reduced the number of propositions substantiating the principles of Geometry, and I have classified them in such a manner that particular cases are enunciated by general propositions, a change which is likely to impress on the pupils the accuracy of geometrical principles, as they will be shown that geometrical principles are the same in all cases and under every circumstance.

Many of the best geometers have objected to the *Reductio ad absurdum* in Geometry, while all geometers prefer the Direct Method of demonstra·

tion. Any *true* proposition is susceptible of being *directly* demonstrated. And without entering into the merits or demerits of the *Reductio ad absurdum*, I have excluded it from geometrical reasoning, and have used the *Direct Method* only, a change which agrees with the spirit of the age, and fulfills the requirements of progress. I have omitted the various diagrams usually put among the definitions of Geometry, because when a magnitude is properly defined, the learner has a better conception of it from the definition than any diagram can give him; and the omission of the diagrams will assist the mental exercise and cultivate the understanding of the learner, which is the great object of geometrical study; and if the learner be made to draw the diagrams from the definitions, he will be better instructed than if they be given by the author. The time is not far distant when geometrical science may be attempted without using diagrams in the demonstrations. The diagrams are auxiliaries to the mind in the ascertainment of truth; they are not necessary to the existence of truth, and " Geometry considers all bodies in a state of *abstraction*, very different from that in which they actually exist, and the truths it discovers and demonstrates are pure abstractions, hypothetical truths." Hence, diagrams are like the pebbles used by Indians in counting, or other means of computing before the principle of numeration was discovered; that when the intellect of man becomes more highly expanded and cultivated, diagrams will be regarded necessary to the first conceptions of geometrical knowledge, but altogether unsuited to a high development of geometrical science.

I am greatly indebted to Hon. S. S. Randall, City Superintendent of the Board of Education of New York, for many valuable suggestions in the demonstrations and present arrangement of this work; and also under many obligations to Professor Docharty, of the College of the City of New York; and to L. D. Gale, M.D., General Secretary of the New York Association for the Advancement of Science and Art, Cooper Union.

LAWRENCE S. BENSON.

61 MORTON STREET, CITY OF NEW YORK,
April 16th, 1867.

THE ELEMENTS OF EUCLID AND LEGENDRE.

BOOK FIRST.

ON THE STRAIGHT LINE AND TRIANGLE.

DEFINITIONS.

1. A *definition* is the precise term by which one thing is distinguished from all other things.

2. *Mathematics* is that science which treats of those abstract quantities known as numbers, symbols, and magnitudes.

3. *Geometry* is that branch of Mathematics where the extensions of magnitudes are considered without regard to the actual existence of those magnitudes.

4. A *magnitude* has one or more of three dimensions, viz., length, breadth, and thickness.

5. Geometers define a *point*, position without magnitude; but to give a point position, would entitle it to the three dimensions of magnitude, whereas a point in Geometry expresses no dimension.

6. A *diagram* represents the abstractions of magnitudes, whereby their dimensions are determined, and geometrical reasoning conducted without regard to the actual properties of those magnitudes.

7. A *line* expresses length only, and is capable of two conditions—it can be straight or curved; when its length is always in one direction, it is straight; but when there is a continual variation in the direction of its length, it is curved, or in brevity called a curve.

Scholium. A straight line can not be defined as having all its points in the same direction, because the points of a line are its extremities, and the extremities of a curved line can be placed on a straight line, and in this case the definition would not distinguish a straight line from a curve. And if a line be regarded composed of points, this would infer that a point has dimension; but the intersection of lines is a point, which, however, does not give position to the point, because a line is an abstraction, and position implies actual existence.

8. A *surface* expresses an inclosure by not less than three straight lines, or by one curved line, or by one straight line and one curved line; consequently a surface has breadth and length, and the extremities of surfaces are lines, and the intersection of ´ one surface with another is a line.

Scho. A *plane* surface, or sometimes called a *plane*, is one in which any line can be drawn wholly in the surface; and a *c•rved* surface is one in which a curve only can be drawn wholly in the surface in the direction of the curvature.

9. A *volume* or *solid* expresses an inclosure made by surfaces, and has breadth, length, and thickness; the extremities of a volume are surfaces, and the intersection of one volume with another is a surface.

10. An *angle* is formed by *two* straight lines meeting each other; the point of intersection of the lines is called the *vertex* of the angle. When one straight line meets another straight line, so as to make two adjacent angles, these angles are *right angles* when they are equal; and when one angle is greater than the other angle, the greater angle is an *obtuse* angle, and the less angle is an *acute* angle. The straight line which makes the two adjacent angles equal is the *perpendicular* to the other straight line.

11. When two straight lines on the same plane never meet each other on whichever side they be produced, they are called *parallel* lines.

12. *Rectilinear* surfaces are contained by straight lines, and are called *polygons ;* when a polygon has three sides, it is a *triangle ;* when it has four sides, it is a *quadrilateral ;* when it has five sides, it is a *pentagon ;* when it has six sides, it is a *hexagon ;* when it has seven sides, it is a *heptagon ;* when it has eight sides, it is an *octagon ;* when it has nine sides, it is an *enneagon ;* when it has ten sides, it is a *decagon ;* and so on, being distinguished by particular names derived from the Greek

language, denoting the number of angles formed by the sides. The straight line drawn through two remote angles of a polygon of four or more sides, is a *diagonal*.

13. When the triangle has its three sides equal, it is *equilateral;* when two of its sides are equal, it is *isosceles;* and when its sides are unequal, it is *scalene.* When its angles are equal, it is *equiangular;* when one of its angles is a right angle, it is *right-angled;* when one of its angles is an obtuse angle, it is *obtuse-angled;* and when all its angles are acute, it is *acute-angled.*

14. When a quadrilateral has its opposite sides parallel, it is a *parallelogram;* when it has two sides only parallel, or none of its sides parallel, it is a *trapezium.*

15. When a parallelogram has a right angle, it is a *rectangle;* when it has two adjacent sides equal, but no right angle, it is a *rhombus.* When the rectangle has its sides equal, it is a *square;* and when its opposite sides only are equal, it is an *oblong.* When the parallelogram has its opposite sides only equal, and no right angle, it is a *rhomboid.*

16. A plane surface contained by one line is a circle when every part of the line is equally distant from a point in the surface; the point is the *center* of the circle, and the line is the *circumference.*

17. The straight line drawn from the center to the circumference is the *radius;* the straight line drawn from one part of the circumference through the center to another part of the circumference is the *diameter,* which divides the circle and circumference each into two equal parts. When the straight line does not pass through the center, it is a *chord.*

18. That portion of the circle contained by the semicircumference and diameter is a *semicircle;* and that portion contained by the chord and a part of the circumference is a *segment;* a part of the circumference is an *arc.*

19. If the vertex of an angle be the center of a circle, that part of the circumference intercepted by the sides of the angle will give the value of the angle; hence, the angle is *measured* by an arc when its vertex is the center of the circle. But when the vertex is in the circumference, the angle is *subtended* by the arc intercepted by its sides; hence, equal angles will be

measured by equal arcs, and subtended by equal arcs; therefore equal arcs measure or subtend equal angles.

20. Two arcs are *supplementary* when both together are equivalent to the semicircumference. And two angles are supplementary when both together are equivalent to two right angles, and *complementary* when equivalent to one right angle.

21. Things are equal when they have equal magnitudes and when they coincide in all respects; and are *equivalent* when they have equal magnitudes, but do not coincide in all respects.

22. The term, *each to each*, or sometimes *respectively*, is a limiting expression, and is used to denote the equality of lines or magnitudes taken in the same order; for without this qualification, two lines or magnitudes said to be equal to two other lines or magnitudes, would imply that their sums are equal, when it would be desirous of meaning that they are equal in the same order in which they are expressed—a difference very important in the demonstration of a proposition.

23. A proposition is demonstrated by *superposition* when one figure is supposed applied to another, which is done in the first case of the third proposition of this book.

24. One proposition is the *converse* of another when, in the language of logic, the subject of the latter is the predicate of the former, and the predicate of the latter is the subject of the former.

METHOD OF REASONING.

1. From the foregoing *definitions*, it is shown that the straight line and curve have certain relations, uses, and properties which are important to be known. And in order that these relations, uses, and properties may be satisfactorily interpreted, there are certain terms, expressive of certain facts or states of knowledge, by means of which the mind intuitively perceives a connection between the things known and those for elucidation, such as *axioms, hypotheses*, and *postulates;* as *demonstrations, theorems, problems*, and *lemmas;* as *corollaries* and *scholiums*. With the assistance of these, the mind is carried step by step in all its investigation of extension, and is able to discover by such investigation the properties, uses, and relations of geometrical magnitudes. They are the *data* by

which the hidden truths are revealed. Upon them a system of logic or argumentation is conducted, and by the conformity of the arguments and conclusions with the accepted truths, we have the science of Geometry.

2. *Proposition* in Geometry is a general term, expressing the subjects to be considered, and is either a problem or theorem. When it is the first, there is something required to be performed, such as drawing a line or constructing a figure; and whatever points, lines, angles, or other magnitudes are given to effect the purpose, they are the *data* of the problem ; and when it is the latter, a truth is proposed for demonstration, and whatever is assumed or admitted to be true, and from which the proof is to be derived, is the *hypothesis*.

3. *Demonstration* consists in evident deductions from clear premises, whereby the conclusion corroborates the premises and shows the argumentativeness of the deductions. In the course of demonstration, reference is often made to some previous proposition or definition.

4. Sometimes inferences arise involving another principle, but do not require any long process of reasoning to establish their truth—these are *corollaries*. Any remark made from the demonstration of a proposition is a *scholium*. A proposition which is preparatory to one or more propositions, and is of no other use, is a *lemma*.

5. And for the establishment of a proposition, there are four things required, viz.: the general enunciation, the particular enunciation, the construction, and the demonstration.

6. The hypotheses of demonstration are known as *axiom* and *postulate ;* the former is assumed to prove the truth of a theorem, and the latter is granted to perform the requisites of a problem.

7. An axiom is so evidently clear, that no process of reasoning can make it more clear; its truth is so easily recognized by the human mind, that so soon as the terms by which it is expressed are understood, it is admitted; for instance, it is assumed as

AXIOMS.

1. Things which are equal to the same, or to equals, are equal to one another.

2. If equals or the same be added to equals, the wholes are equal.

3. If equals or the same be taken from equals, the remainders are equal.

4. If equals or the same be added to unequals, the wholes are unequal.

5. If equals or the same be taken from unequals, the remainders are unequal.

6. Things which are doubles of the same, or of equals, are equal to one another.

7. Things which are halves of the same, or of equals, are equal to one another.

8. Magnitudes which exactly coincide with one another are equal.

9. The whole is greater than its part.

10. The whole is equal to all its parts taken together.

11. All right angles are equal to one another.

12. If a straight line meet two other straight lines which are in the same plane, so as to make the two interior angles on the same side of it, taken together, less than two right angles, these straight lines shall at length meet upon that side, if they be continually produced.

These are the *self-evident* truths used by Euclid for geometrical demonstration; but if the first *eleven* be considered for awhile, it will be seen that they can be reduced to *two* general axioms, viz., things which are equal to the same are equal, and things which are not equal to the same are unequal; because when we add, subtract, multiply, or divide equals, the equality in each case is not destroyed; hence in each case equal to one another. And when we add unequals to or subtract unequals from equals, the sums or remainders are not equal to the same, hence unequal to one another. And magnitudes which exactly coincide with one another are equal to the same, hence equal; a whole and a part are not equal to the same, hence are unequal; while a whole and all its parts are equal to the same, hence are equal. From the definition of right angles, it is seen that when a straight line meets another straight line, so as to make the two adjacent angles formed by them equal to one

another, the two adjacent angles are right angles; then these
two right angles are equal; and since *all right angles* agree with
the definition, they are equal to the same thing, hence equal to
one another. But the twelfth axiom is not self-evident, because the converse has been demonstrated, viz.: that two
straight lines which meet one another make with any third line
the interior angles less than two right angles. Geometers perceiv-
ing this blemish in the Elements of Euclid, have endeavored in
many ways to remove it, but without complete success. They
employed three methods for this purpose: 1. By adopting a
new definition of parallel lines. 2. By introducing a new
axiom. 3. By reasoning from the definition of parallel lines,
and the properties of lines already demonstrated.* The difficulty with parallel lines is, that geometers have confounded a
definition with a proposition. Definition 11 is perfectly legiti-
mate, as it simply defines what kind of lines are parallel; but
when it is inferred from it that these lines are equally distant
from each other, this is no axiomic inference, because the curve
and its asymptote are two lines which never meet, however far
they be produced on the same plane, but they are not equally
distant from each other; hence the inference that parallel lines
are equally distant, embodies a question which requires a demonstration to establish; and to establish this question has given
perplexity to geometers, for though they have proven the lines
equally distant at particular points, they have not proven them
so at every point; and here consists the incompleteness of their
demonstrations, and here is required some general demonstration which will embrace every part of the lines, however so far
they be produced on the same plane.†

8. A postulate is a problem so easy to perform that it does
not require any explanation of the manner of doing it, so that
the geometer reasonably expects the method to be known; for
instance, it is granted as—

---

* See notes to Playfair's Euclid, Legendre's Geometry, Leslie's Geom-
etry, the *excursus* to the first book of Camerer's Euclid, Berlin, 1825;
Col. P. Thomson's *Geometry without Axioms*, Professors Thomson's and
Simson's editions of Euclid—London, Glasgow, and Belfast.

† See fifteenth and nineteenth propositions of this book.

1. That a straight line can be drawn from any one point to any other point.

2. That a terminated straight line may be extended to any length in a straight line.

3. That a circle may be described from any center, at any distance from that center.

EXPLANATION OF SIGNS.

In Algebra, the sign +, called *Plus* (*more by*), placed between the names of two magnitudes, is used to denote that these magnitudes are added together; and the sign —, called *Minus* (*less by*), placed between them, to signify that the latter is taken from the former. The sign =, which is read *equal to*, signifies that the quantities between which it stands are equal to one another. The sign ⌒⌒, signifies that the quantities between which it stands are *equivalent to* one another.

In the references, the Roman numerals denote the book, and the others, when no word is annexed to them, indicate the proposition; otherwise the latter denote a definition, postulate, or axiom, as specified. Thus, III. 16 means the sixteenth proposition of the third book; and I. ax. 2, the second axiom of the first book. So also *hyp.* denotes *hypothesis*, and *const.* *construction*.

PROPOSITIONS.

PROP. I.—PROBLEM.—*To describe an isosceles triangle on a finite straight line given in position.*

Let AX be the given straight line; it is required to describe an isosceles triangle having its base on AX.

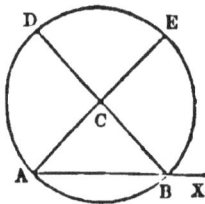

From a point C, without the line AX as a center, and a radius CA (I. post. 3), describe a circle ABED, cutting the line AX in two points A and B; draw from these points the straight lines AE and BD (I. post. 1) passing through the center of the circle; the triangle ACB is the one required.

Because C is the center of the circle ABED (I. def. 16), CA

is equal to CB, therefore the triangle ACB has two sides equal; hence (I. def. 13) it is isosceles, and is described on AX, which was required to be done.

Corollary 1. But the angle EAB is subtended by the arc EB (I. def. 19), and the angle DBA is subtended by the arc DA; since AE and DB pass through the center of the circle (const.), they are both diameters of the circle (I. def. 17); hence the arcs DEB and ADE are each a semicircumference, and (I. ax. 1) are equal; therefore the sum of the arcs BE and ED is equivalent to the sum of the arcs ED and DA; the arc ED is common; hence (I. ax. 3) we have the arc EB equal to the arc DA; therefore the angles EAB and DBA are subtended by equal arcs, consequently (I. def. 19) the angles are equal. Hence, in an isosceles triangle, the angles opposite the equal sides are equal.

Cor. 2. The line AE, which forms with AB the angle EAB, intercepts the line DB at C, which forms with AB the angle DBA, and the line DB intercepts AE at C also. C being the center of the circle ABED, CB, that portion of BD intercepted by AE, is equal to CA, that portion of AE intercepted by BD (I. def. 16); but CB and CA are the sides of the triangle ACB (I. 1); hence, when two angles of a triangle are equal, the opposite sides to them are also equal, and the triangle is isosceles (I. def. 13).

PROP. II.—PROBLEM.—*To describe an equilateral triangle on a finite straight line given in magnitude.*

Let AB be the given straight line; it is required to describe an equilateral triangle having AB for its base.

From A as a center, and a radius AB (I. post. 3), describe the circle FCD; and from B as a center, and a radius BA, describe the circle HCE. The circles having equal radii (I. ax. 1) are equal; draw from C through the center B, CH; and from C through the center A, CF (I. post 1); the triangle ACB is the one required.

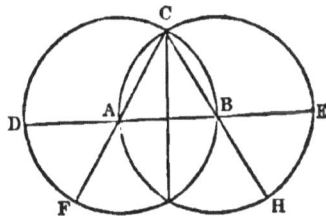

Because the circles have equal radii (const.), AC is equal to

AB, and CB is equal to AB; hence (I. ax. 1) the three sides of the triangle ACB are equal; the triangle (I. def. 13) is equilateral, and is described on AB, which was required to be done.

Corollary 1. If AB be produced both ways (I. post. 2) to D and E, the angle DBC is subtended by the arc CD, and the angle FCB is subtended by the arc FB; the arcs CB and BF are together equivalent to the arcs BC and CD (I. ax. 1); hence (I. 1, cor. 1) the arc BF is equal to the arc CD, therefore (I. def. 19) the angle FCB is equal to the angle DCB. Again: the angle ACH is subtended by the arc AH, and the angle CAE is subtended by the arc CE. But (I. ax. 1) the arcs AC and CE are equivalent to the arcs CA and AH, and (I. ax. 3) the arcs CE and AH are equal; therefore (I. def. 19) the angles ACH and CAE are equal, but the angle ACH is the same as the angle FCB; hence (I. ax. 1) the three angles of the triangle are equal; therefore in an equilateral triangle the angles are equal. And in a manner similar to Cor. 2 of the first proposition, it can be shown, *conversely*, that when a triangle has three equal angles, the sides opposite them are also equal; hence an equilateral triangle is also equiangular, and, *conversely*, an equiangular triangle is also equilateral.

PROP. III.—THEOREM.—*If two triangles have two sides of the one equal to two sides of the other, each to each, and have also an angle in one equal to an angle in the other similiarly situated with respect to those sides, the triangles have their bases or remaining sides equal; their other angles equal, each to each, viz., those to which the equal sides are opposite, and the triangles are equal.*

This general proposition has four cases, viz.: first, when the equal angles are contained by the respectively equal sides; second, when the equal angles are opposite to one pair of the respectively equal sides; third, when the equal angles are opposite to the other pair of the respectively equal sides; and fourth, the limitation that when the least sides respectively of the triangles be equal, and the angles opposite the least sides be equal, the angles opposite the greater of the respectively equal sides must be of the same kind, either both acute, or not acute.

First case.   Let ABC and DEF be the two triangles having
any two sides equal, each to each,
viz., AC and CB equal to EF and
DF, and the contained angles ACB
and EFD equal; the remaining sides
AB and DE are equal, the angle CBA
opposite AC equal to the angle FDE
opposite FE, the angle CAB opposite
CB equal to the angle FED opposite
DF, and the triangles ABC and DEF
are equal.

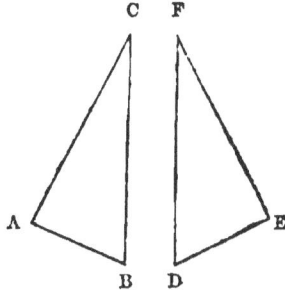

If the triangle ABC be placed on the triangle DEF so that
the vertex of the angle ACB will fall on the vertex of the
angle DFE, the angle ACB being equal to the angle DFE
(hyp.), the side CB will fall on FD, and the side CA will fall on
FE; CB and FD being equal (hyp.), the extremity B will fall
on the extremity D.   CA and FE being equal (hyp.), the extremity A will fall on the extremity E; and since AB is a
straight line, it will coincide with DE (I. def. 7), a straight line
drawn from D to E.   Therefore the triangle ABC has its three
sides coinciding with the three sides of the triangle DEF;
hence the angle CAB will fall on the angle FED, and be equal
to it; the angle CBA will fall on the angle FDE, and be equal
to it; consequently the two triangles have their three sides and
three angles equal, each to each, and (I. ax. 8) are equal.

Second case.   When the triangles ABC and DEF have the
sides CA and CB respectively equal to FE and FD, and the
angles ABC and EDF equal, respectively opposite to CA and
FE, the remaining sides are equal; the angle CAB opposite CB
is equal to the angle FED opposite to FD, the angle ACB opposite to AB is equal to the angle EFD opposite to DE, and
the triangles are equal.

Let the side DE
be put on AB so
that D will fall
on B, and the
equal angles ABC
and EDF will be
on different sides of AB; join CF (I. post. 1).   BC and BF (hyp.) are

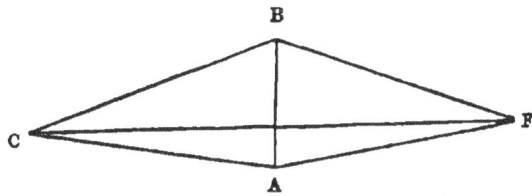

equal,the triangle CBF (I. def. 13) is isosceles, and (I. 1, cor. 1) the angle BCF equal to the angle BFC; and because CA and AF are equal (hyp.), the triangle CAF (I. def. 13) is isosceles, and (I. 1, cor. 1) the angle ACF is equal to the angle AFC; then (I. ax. 2) the angles BCF and ACF are equal to the angles BFC and AFC, or the angle BCA equal to the angle BFA (I. ax. 1 and ax. 10). Hence we have in the triangles ABC and ABF, two sides, and the contained angle in each equal, each to each, therefore by *first case* the triangles can be shown equal in all respects.

Third case. It can be proven in a similar manner as the second case. When the angles CAB and FED of the second case are obtuse, and the angles CBA and FDE of the third case are obtuse, the proofs are given by the third axiom of the first book.

Fourth case. When the triangles ABC and DEF have their least sides in each equal—viz., AB to DE—and another side in each, equal, the angles ACB and EFD being equal, the angles opposite the second pair of equal sides must be both acute or both not acute ; otherwise, two triangles can be formed having two sides and an angle in each equal, each to each, and the triangles unequal. For, in the triangles ABC and ACD, the side AC is common, the angle BAC equal to angle CAD, the sides BC and CD can be equal, and the triangles (I. ax. 9) unequal; hence in two triangles when the greatest and least sides are respectively equal, and the equal angles opposite to the least sides be given, the angles opposite the greatest sides must both be not acute to determine the triangles ; but when in two triangles the two less sides of each are respectively equal, and the equal angles opposite the least sides be given, the angles opposite the other equal sides must both be acute, to determine the triangles.

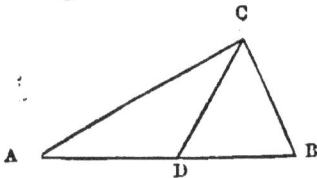

The equality of the triangles can be proven by the second and third cases, using the second axiom when the angles are acute, and the third axiom when the angles are obtuse; but when the angles are right-angled, the equality of the triangles is shown from the first corollary to the first proposition without those axioms.

PROP. IV.—THEOR.—*If the three sides of one triangle be equal to the three sides of another, each to each:* (1) *the angles of one triangle are equal to the angles of the other, each to each, viz., those to which the equal sides are opposite, and* (2) *the triangles are equal.*

Let ABC and DEF be the two triangles having their three sides equal,
viz., AB to DE, CA to FE, and CB
to FD, the angles are equal, viz., ACB
to EFD, CAB to FED, and CBA
to FDE; and the triangles are equal.

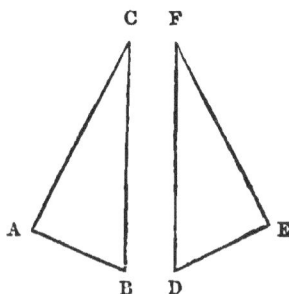

If the side DE be placed on the
side AB so that the triangles will
fall on different sides of AB, D will
fall on B, and E on A, because DE is equal to AB, and the
triangle DEF will
take the position
BFA, BF being
the same as DF,
and FA the same
as FE. Join CF,
and because (hyp.) BC is equal to BF, the angles BCF and
BFC are equal (I. 1, cor. 1). It would be shown in a similar
manner that the angles FCA and CFA are equal, therefore (I.
ax. 2) the angles BCA and BFA are equal—that is (I. ax. 1), the
angles BCA and DFE are equal. But (hyp.) the sides CB and
FD are equal, and the sides CA and FE, and it has been shown
that the contained angles are equal, therefore (I. 3, first case)
the other angles are equal—that is, CAB to FED and CBA to
FDE, and the triangles are equal. Wherefore, if the three
sides, etc.

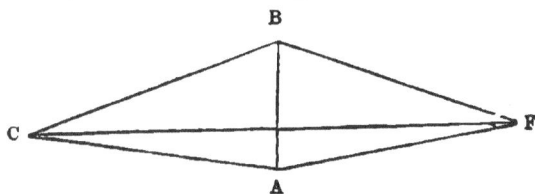

PROP. V.—PROB.—*To bisect a given angle, that is, to divide it into two equal angles.*

Let BAC be the given angle; it is required to bisect it.

From A as a center, and AD less than AB (I. post. 3), describe the arc DE; draw the chord DE, then upon DE, on the
side remote from A, describe an equilateral triangle (I. 2), DFE,
then join AF; AF bisects the angle BAC.

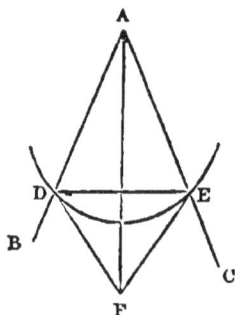

Because AD is equal to AE (L def. 16), and AF is common to the two triangles DAF and EAF, the two sides DF and EF are equal (I. 2); therefore the two triangles DAF and EAF have their three sides equal, each to each, and the triangles are equal (I. 4); consequently the angle DAF opposite DF is equal to the angle EAF opposite EF, and the angle BAC is bisected by the line AF; which was to be done.

OTHERWISE,

Let BAC be the given angle; in AB take any two points as B and D, and cut off AC and AE respectively equal to AB and AD, join BE and CD, and the straight line joining the intersection of BE and CD with the vertex A bisects BAC. The proof is easy, and is omitted to exercise the ingenuity of the pupil.

Prop. VI.—Prob.—*To bisect a given finite straight line.*

Let AB be the given line; it is required to bisect it.

Describe (I. 2) upon it an equilateral triangle ABC, and bisect (I. 5) the angle ACB by the straight line CD; AB is bisected in the point D.

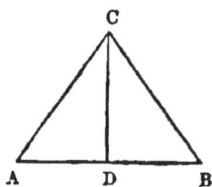

Because AC is equal to CB, and CD common to the two triangles ACD, BCD, the two sides AC, CD are equal to BC, CD, each to each; and the angle ACD is equal (const.) to the angle BCD; therefore the base AD is equal (I. 3) to the base DB, and the line AB is bisected in the point D; which was to be done.

Scho. In practice, the construction is effected more easily by describing arcs on both sides of AB, from A as a center, and with any radius greater than the half of AB; and then, by describing arcs intersecting them, with an equal radius, from B as center, the line joining the two points of intersection will bisect AB. The proof is easy.

Prop. VII.—Prob.—*To draw a straight line perpendicular*

*to a given straight line, from a given point in that straight
line.*

Let AB be the given straight line, and C a point given in it;
it is required to draw a perpendicular from the point C.

From C as a center, and a radius
CE (I. post 3), describe the semicircle
EHF; then (I. def. 16) EC is equal to
CF, and on EF (I. 2) describe the
equilateral triangle EDF; then a line
from C to the vertex D is the perpendicular required.

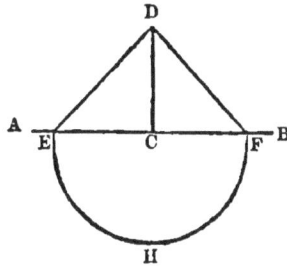

Because EC is equal to CF (I. def.
16), ED is equal to DF (I. 2), and the angle DEC equal to
angle DFC (I. 2, cor. 1); hence (I. 3) the triangles ECD and
FCD are equal; but the angle ECD is equal to the angle FCD,
and therefore (I. def. 10) DC is perpendicular to AB from C.

PROP. VIII.—PROB.—*To draw a straight line perpendicular
to a given straight line of an unlimited length, from a given
point without it.*

Let AB be the given straight line, which may be produced
any length both ways, and let C be a point without it. It is
required to draw a straight line from C
perpendicular to AB.

Take any point D upon the other side
of AB, and from the center C, at the distance CD, describe (I. post. 3) the circle
ADB meeting AB in A and B; bisect (I. 6) AB in G, and join
CG; the straight line CG is the perpendicular required.

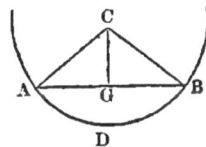

Join CA, CB. Then, because AG is equal to GB, and CG
common to the triangles AGC, BGC, the two sides AG, GC
are equal to the two, BG, GC, each to each; and the base CA
is equal (I. def. 16) to the base CB; therefore the angle CGA
is equal (I. 4) to the angle CGB; and they are adjacent angles;
therefore CG is perpendicular (I. def. 10) to AB. Hence, from
the given point C a perpendicular CG has been drawn to the
given line AB; which was to be done.

Scho. This proposition and the preceding contain the only
two distinct cases of drawing a perpendicular to a given straight

line through a given point; the first, when the point is *in* the line; the second, when it is *without* it.

In practice, the construction will be made rather more simple by describing from A and B, when found, arcs on the remote side of AB from C, with any radius greater than the half of AB, and joining their point of intersection with C.

PROP. IX.—THEOR.— *When one straight line meets another straight line and forms two unequal angles on the same side of that line, the two angles will be equivalent to two right angles.*

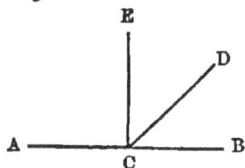

Let the straight line DC meet the straight line AB and form the two un-equal angles DCA and DCB on the same side of AB; the two angles will be equiva-lent to two right angles.

At C, where the line DC meets AB, draw a perpendicular to AB from C (I. 7), then the angles ACE and ECB are two right angles (I. def. 10). But (I. ax. 10) the angle ECB is equivalent to the angles ECD and DCB both together; likewise the angles ACE and ECB together are equivalent to the angles ACD and DCB both together; hence (I. ax. 1) the angles ACD and DCB together are equivalent to two right angles. Wherefore, when one straight line meets, etc.

Cor. Hence, if the straight line DC be produced on the other side of AB, the four angles made by DC produced and AB are together equivalent to four right angles.

Hence, also, all the angles formed by any number of straight lines intersecting one another in a common point are together equivalent to four right angles.

PROP. X.—THEOR.—*If, at a point in a straight line, two other straight lines on the opposite sides make the adjacent angles together equivalent to two right angles, those two straight lines are in one and the same straight line.*

Let DC be the straight line which makes, at the point C, with AC and CB, two adjacent angles ACD and DCB together equivalent to two right angles; AC and CB are in one and the same straight line.

From C draw a perpendicular to AC (I.
7), then the angle ACE is a right angle
(I. def. 10). But (hyp.) ACD and DCB
are together equivalent to two right
angles, so ACE and ECB are equivalent
to two right angles (I. ax. 1); hence ACE being a right angle
(const. and I. def. 10), ECB must also be a right angle; then
EC is perpendicular to CB (I. def. 10), and the angles ACE and
ECB are equal (I. ax. 11); therefore (I. def. 10) EC is a straight
line which makes two equal angles with AB. But AC and CB
make with EC the same equal angles; hence AC and CB are
the same straight line with AB. And DC makes with AC and
CB (hyp.) two adjacent angles equivalent to two right angles,
but DC makes with AB (I. 9) the same angles equivalent to
two right angles; hence AC and CB are the same straight line
with AB (I. def. 7).

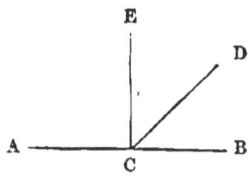

OTHERWISE,

It is proven (I. 9) that the angles ACD and DCB are together equivalent to two right angles; then, as C is a point in
AB, AC can be one line and CB another (I. def. 7, scho.);
hence AC and CB are in one and the same straight line, because the two unequal angles ACD and DCB are equivalent to
two right angles (I. 9). Wherefore if, at a point, etc.

PROP. XI.—THEOR.—*If two straight lines cut one another,
the vertical or opposite angles are equal.*

Let the two straight lines AB, CD cut one another in E; the
angle AEC is equal to the angle DEB, and CEB to AED.

Because the straight line AE makes with CD the angles
CEA, AED, these angles are together equivalent (I. 9) to two
right angles. Again: because DE
makes with AB the angles AED, DEB,
these also are together equivalent to
two right angles; and CEA, AED
have been demonstrated to be equivalent to two right angles;
wherefore (I. ax. 11 and 1) the angles CEA, AED are equal to
the angles AED, DEB. Take away the common angle AED,
and (I. ax. 3) the remaining angles CEA, DEB are equal.

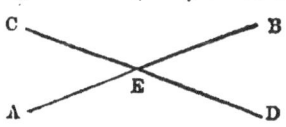

In the same manner it can be demonstrated that the angles CEB, AED are equal. Therefore, if two straight lines, etc.

Cor. If at a point in a straight line two other straight lines meet on the opposite sides of it, and make equal angles with the parts of it on opposite sides of the point, the two straight lines are in one and the same straight line.

Let AEB be a straight line, and let the angles AEC, BED be equal, CE, ED are in the same straight line. For, by adding the angle CEB to the equal angles AEC, BED, we have BED, BEC together equal to AEC, CEB, that is (I. 9), to two right angles; and therefore, by this proposition, CE, ED are in the same straight line.

Scho. In the proof here given, the common angle is AED; and CEB might with equal propriety be made the common angle. In like manner, in proving the equality of CEB and AED, either AEC or BED may be made the common angle. It is also evident, that when AEC and BED have been proved to be equal, the equality of AED and BEC might be inferred from the ninth proposition, and the third axiom.

PROP. XII.—PROB.—*To describe a triangle of which the sides shall be equal to three given straight lines ; but any two of these must be greater than the third.*

Let A, B, C be three given straight lines, of which any two are greater than the third; it is required to make a triangle of which the sides shall be equal to A, B, C, each to each.

Take an unlimited straight line DE, and let F be a point in it, and make FG equal to A, FH to B, and HK to C. From the center F, at the distance FG, describe (I. post. 3) the circle GLM, and from the center H, at the distance HK, describe the circle KLM. Now, because (hyp.) FK is greater than FG, the circumference of the circle GLM will cut FE between F and K, and therefore the circle KLM can not lie wholly within the circle GLM. In like manner, because (hyp.) GH is greater than HK, the circle GLM can not lie wholly within the circle KLM. Neither can the circles be

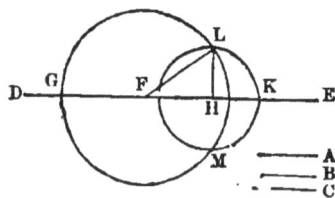

wholly without each other, since (hyp.) GF and HK are together greater than FH. The circles must therefore intersect each other; let them intersect in the point L, and join LF, LH; the triangle LFH has its sides equal respectively to the three lines A, B, C.

Because F is the center of the circle GLM, FL is equal (I. def. 16) to FG; but (const.) FG is equal to A; therefore (I. ax. 1) FL is equal to A. In like manner it may be shown that HL is equal to C, and (const.) FH is equal to B; therefore the three straight lines LF, FH, HL are respectively equal to the three lines A, B, C; and therefore the triangle LFH has been constructed, having its three sides equal to the three given lines, A, B, C; which was to be done.

Scho. It is evident that if MF, MH were joined, another triangle would be formed, having its sides equal to A, B, C. It is also obvious that in the practical construction of this problem, it is only necessary to take with the compasses FH equal to B, and then, the compasses being opened successively to the lengths of A and C, to describe circles or arcs from F and H as centers, intersecting in L; and lastly to join LF, LH.

The construction in the proposition is made somewhat different from that given in Simson's Euclid, with a view to obviate objections arising from the application of this proposition in the one that follows it.

PROP. XIII.—PROB.—*At a given point in a given straight line, to make a rectilineal angle equal to a given one.*

Let AB be the given straight line, A the given point in it, and C the given angle; it is required to make an angle at A, in the straight line AB, that shall be equal to C.

In the lines containing the angle C, take any points D, E, and join them, and make (I. 12) the triangle AFG, the sides of which, AF, AG, FG, shall be equal to the three straight lines CD, CE, DE, each to each. Then, because FA, AG are equal to DC, CE, each to each, and the base FG to the base DE, the angle A is equal (I.

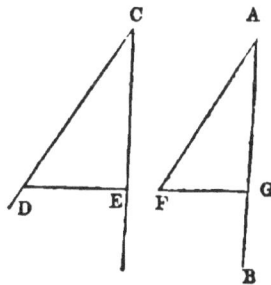

4) to the angle C. Therefore, at the given point A in the given straight line AB, the angle A is made equal to the given angle C; which was to be done.

Scho. The construction is easy, by making the triangles isosceles. In doing this, arcs are described with equal radii from C and A as centers, and their chords are made equal.

It is evident that another angle might be made at A, on the other side of AB, equal to C.

PROP. XIV.—THEOR.—*If two angles of one triangle be equal to two angles of another, each to each, and if a side of the one be equal to a side of the other similarly situated with respect to those angles ; (1) the remaining sides are equal, each to each ; (2) the remaining angles are equal ; and (3) the triangles are equal.*

This proposition is the converse of the third proposition, and is susceptible of three cases, viz.: first, when the equal sides are between the equal angles ; secondly and thirdly, when the equal sides are opposite to the equal angles similarly situated.

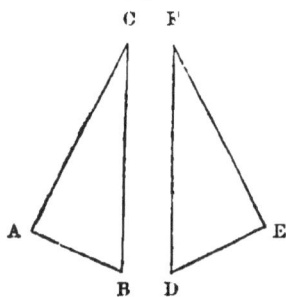

Let ABC and DEF be two triangles which have the angles ACB and EFD equal; the angles BAC and DEF equal, and the sides CA and FE equal; then the sides CB and FD are equal, also the sides AB and DE; the angles CBA and FDE are equal, and the triangles are equal to one another.

If the triangles be placed so as to have their sides CB and FD in the same straight line, but the triangles be on opposite sides of that line, and the vertex C on the vertex F; then because the angles AFD and EFD are equal (hyp.), the angle AFE is bisected by FD; and because AF and FE are equal (hyp.), the triangle AFE is isosceles (I. def. 13), and the line AE (I. 6) is also bisected by FD; hence (I. 1, cor. 1) the angles FAE and FEA are equal; therefore (I. 3) the triangles FAH and FEH are equal. But the angles FAD

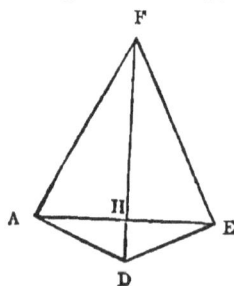

and FED are equal (hyp.); then taking from each the equal
angles FAH and FEH, there will remain (I. ax. 3) the angle
HAD equal to the angle HED; hence (I. 1, cor. 2) the triangle
AED is isosceles, and (I. def. 13 and I. cor. 2) the side AD is
equal to the side ED, and AE being bisected by FD (I. 6), we
have the triangles AHD and EHD (I. 3) equal; therefore the
angles HDA and HDE are equal; hence the triangles AFD
and EFD have the sides AF and EF (hyp.) equal, the angles
FAD and FED equal (hyp.), and the sides AD and DE equal
(I. def. 13 and I. cor. 2); therefore (I. 3) the angles FDA and
FDE are equal; the side FD is common, and the two triangles
are equal.

In a similar manner, the second and third cases can be demonstrated. Wherefore, if two angles of one triangle be, etc.

Cor. Hence, from this proposition, it can also be shown that
the second corollary to the first proposition is true. Let the
triangle ABC have the angle CAB equal
to the angle CBA, then will AC be equal
to CB.

From the vertex C (I. 8) draw CD perpendicular to AB, then the angles CDA
and CDB are both right angles (I. def. 10),
the angles CAD and CBD are equal (hyp.), and the side CD common (const.); therefore (I. 14) the sides AC and CB are equal.

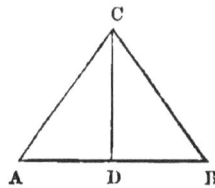

Scho. It will be seen from propositions third, fourth, and
fourteenth of this book, that two triangles are in every respect
equal when the three sides of the one are respectively equal to
the three sides of the other, when two triangles have two angles
and a side in each equal, each to each, and when one angle and
two sides of one are equal to one angle and two sides of the
other, each to each, and the equal angles in each triangle similarly situated with respect to those sides, but with this limitation, that when two equal sides respectively of the triangles are
the least sides, that the angles opposite the greater sides respectively of the triangles must be of the same kind, either both
acute, both right-angled, or both obtuse. And from these propositions it is shown that of the sides and angles of a triangle,
three must be given to determine the triangle, and these three
can not all be angles. Were only three angles given, the sides,

as will appear from the twentieth prosposition of this book, might be of any magnitude whatever. The pupil may occupy himself in proving these propositions by *superposition* or some other way, as by pursuing a course of demonstration different from what is given in the text, he will more readily familiarize himself with the process of geometrical reasoning.

PROP. XV.—THEOR.—*Parallel straight lines are equally distant from each other, however so far they be produced on the same plane.*

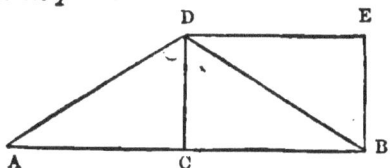

Let the straight line AB be bisected (I. 6) at C, and let the perpendicular CD be drawn (I. 7); join AD and BD, and if the triangle ADC be applied to the triangle BDC so that they will fall on different sides of BD and have D and BD common, their sides DE and CB will be equally distant from each other, and so they will never meet, however so far they be produced on the same plane, and consequently (I. def. 11) are parallel straight lines.

Because DC is perpendicular to AB (const.), the angles ACD and BCD are both right angles (I. def. 10), and are equal (I. ax. 11); hence the triangles ADC and BDC have the side DC common, the sides AC and CB equal (const.), and the angles ACD and BCD equal (I. def. 10 and ax. 11); therefore (I. 3) the triangles are equal, having the sides AD and BD equal; the angle ADC equal to the angle BDC, and the angle DAC equal to the angle DBC. Now, when ADC is applied to BDC so that they will fall on different sides of BD, and have D and BD common (hyp.), since AD is equal to BD, the point A will fall on B; hence the angle EBD will be the same as ADC, and equal to BDC, the angle BED the same as ACD, and equal to DCB, the angle EDB the same as DAC, and equal to DBC, the side DE the same as AC, and equal to AC; the side EB equal to DC; therefore the triangles BDC and BDE are equal (I. ax. 8), having their sides and angles equal, each to each.

Since DE and CB are straight lines, they have no variation in the direction of their lengths from D to E or from C to B (I. def. 7); and because DC is equal to EB, DE and CB are equally distant from each other at their extremities, and having no variation in the direction of their lengths from D to E and from C to B (I. def. 7), they are also equally distant from each other at every part between D and E and C and B, each to each; therefore DE and CB (I. post. 2) on being produced to any length are still the same straight lines, and will have no variation in the direction of their lengths (I. def. 7), consequently they will always be the distant DC or EB from each other, at every part, each to each; and being always the same distant DC or EB from each other, will never meet, and are parallel straight lines (I. def. 11). Wherefore, parallel straight lines, etc.

Cor. 1. In like manner, it can be shown that DC and EB are parallel lines; hence DCBE is a parallelogram (I. def. 14); and since the angles DCB and DEB are equal, and the angles EDB and BDC equal to the angles EBD and DBC (I. ax. 2), and the sides DE and CB equal, also the sides CD and BE equal, a parallelogram has its opposite sides and opposite angles equal.

Cor. 2. Hence parallel lines, DE and CB, intercepted by a straight line, DB, make the alternate angles EDB and CBD equal; and, *conversely*, when the alternate angles EDB and CBD are equal, the lines DE and CB are parallel.

Cor. 3. In the parallelogram, the opposite sides being equal, the straight lines which join the extremities of two equal and parallel straight lines toward the same parts—that is, the nearest extremities together—are themselves equal and parallel; hence a quadrilateral which has two sides equal and parallel is (I. def. 14) a parallelogram.

Cor. 4. Because the triangles DCB and DEB are equal, a diagonal, DB, bisects the parallelogram; and if two parallelograms have an angle of the one equal to an angle of the other, and the sides containing those equal angles respectively equal, the parallelograms are equal, as the parallelograms can be bisected by diagonals subtended by the equal angles, and the triangles thus formed are equal (I. 3); hence (I. ax. 6) the parallelograms are equal; hence, also, if a parallelogram and a triangle be upon the same or equal bases, and between the same parallels, the parallelogram is double the triangle.

Cor. 5. Hence, also, parallelograms upon the same or equal bases, and between the same parallels, are equal ; and triangles upon the same or equal bases and between the same parallels, are equal.

Cor. 6. Hence, from the preceding corollary, it is plain that triangles or parallelograms between the same parallels, but upon unequal bases, are unequal.

Cor. 7. And a straight line drawn from the vertex of a triangle to the point of bisection of the base, bisects the triangle ; and if two triangles have two sides of the one respectively equal to two sides of the other, and the contained angles supplemental (I. def. 20), the triangles are equivalent ; the *converse* is also true.

Cor. 8. If through any point in either diagonal of a parallelogram straight lines be drawn parallel to the sides of the four parallelograms thus formed, those through which the diagonal does not pass, and which are called the complements of the other two, are equivalent.

PROP. XVI.—THEOR.—*If a straight line fall upon two parallel straight lines,* (1) *it makes the alternate angles equal to one another ;* (2) *the exterior angle equal to the interior and remote upon the same side, and* (3) *the two interior angles upon the same side together, equivalent to two right angles.*

Let the straight lines AB, CD be parallel, and let EF fall upon them ; then (1) the alternate angles AGH, GHD are equal to one another ; (2) the exterior angle EGB is equal to the interior and remote upon the same side, GHD ; and (3) the two interior angles BGH, GHD upon the same side are together equivalent to two right angles.

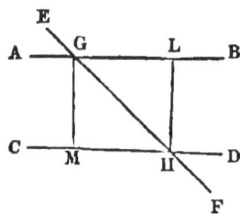

Since AB, CD are parallel (hyp.), the perpendiculars (I. 7) MG, LH make the angles MGH, GHL equal to one another (I. 15, cor. 2), and AGM, LHD are two right angles (I. def. 10) ; if we add AGM to MGH they will be equal to AGH (I. ax. 10), and DHL added to LHG are likewise equal to GHD ; hence (I. ax. 2) AGH and GHD are equal to one another.

Second. AGH is equal to EGB (I. 11), therefore (I. ax. 1) EGB is equal to GHD.

Third. Add to EGB and GHD, each, the angle BGH; therefore (I. ax. 2 and ax. 10) EGB and BGH are equivalent to the angles GHD and BGH, but EGB and BGH are equivalent to two right angles (I. 9); therefore, also, BGH and GHD are together equivalent to two right angles. Wherefore, if a straight line, etc.

Cor. 1. Hence, *conversely*, two straight lines are parallel to one another, if another straight line falling on them (1) makes the alternate angles equal; (2) the exterior angle equal to the interior and remote upon the same side of that line; and (3) the two interior angles upon the same side together equivalent to two right angles.

Let EF fall on AB and CD, the perpendiculars (I. 7) GM and HL make two right angles (I. def. 10), AGM and DHL. But AGH and DHG are equal (hyp.); hence (I. ax. 3) MGH is equal to LHG. But MGL and LHM are both right angles (const. and I. def. 10); hence (I. ax. 3) HGL is equal to GHM; therefore in the triangles HMG and GLH we have two angles in one equal to two angles in the other, each to each, and the side GH common, and (I. 14) the triangles are equal; hence GM and HL are equal, and (I. 15) AB and CD are equally distant from each other, and will never meet on being produced (I. post. 2), and are parallel (I. def. 11).

Because EGB is equal to DHG (hyp.), and EGB equal to AGH (I. 11), the angle AGH is equal to DHG (I. ax. 1), but they are alternate angles; therefore (I. 16, cor. 1, part 1) AB is parallel to CD. Again: because BGH and GHD are together equivalent to two right angles (hyp.), and AGH and BGH are also equivalent to two right angles (I. 9), the angles AGH and BGH are together equal to BGH and GHD (I. 7 and ax. 1); then (I. ax. 3) AGH is equal to GHD, but they are alternate angles, therefore (I. 16, cor. 1, part 1) AB and ED are parallel. Wherefore, two straight lines are parallel, etc.

Cor. 2. When one angle of a parallelogram is a right angle, all the other angles are right angles; for since (I. 16) BGM and GMH are together equivalent to two right angles, if one of them be a right angle, the other must also be a right angle, and

(I. 15, cor. 1) the opposite angles are equal.   A rectangle, then
(I. def. 15), has all its angles right angles.

Cor. 3. If two straight lines make an angle, two others parallel to them contain an equal or supplemental angle; thus LGM, and the vertical angle produced by AB and the continuance of MG through G, are each equal to LHM, while the angle AGM, and its vertical angle contained by GB and the continuance of MG, are each equal to the supplement of LHM; hence we can divide a given straight line AB into any proposed number of equal parts.

PROP. XVII.—THEOR.—*Two straight lines which are not in the same straight line, and which are parallel to a third straight line, are parallel to one another.*

Let the straight lines AB, CD be each of them parallel to the straight line EF; AB is also parallel to CD.

Let the straight line LH cut AB, CD, EF; and because LH cuts the parallel straight lines AB, EF, the angle LGB is equal (I. 16, part 2) to the angle LHF.   Again: because the straight line LH cuts the parallel straight lines CD, EF, the angle LKD is equal (I. 16, part 2) to the angle LHF; and it has been shown that the angle LGB is equal to LHF; wherefore, also, LGB is equal (I. ax. 1) to LKD, the interior and remote angle on the same side of LH; therefore AB is parallel (I. 16, part 1) to CD.   Wherefore, two straight lines, etc.

PROP. XVIII.—PROB.—*To draw a straight line parallel to a given straight line through a given point without it.*

Let AB be the given straight line, and C the given point; it is required to draw a straight line through C, parallel to AB.

In AB take any point D, and join CD; at the point C, in the straight line CD, make (I. 13) the angle DCE equal to CDB; and produce the straight line EC to any point F.

Because (const.) the straight line CD, which meets the two straight lines AB, EF, makes the alternate

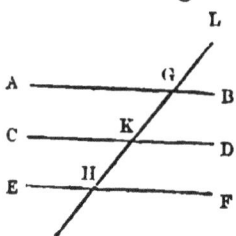

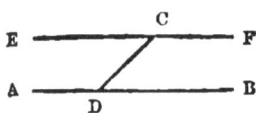

angles ECD, CDB equal to one another, EF is parallel (I. 15, cor. 2) to AB. Therefore the straight line ECF is drawn through the given point C parallel to the given straight line AB; which was to be done.

Prop. XIX.—Theor.—*If a straight line meet two other straight lines which are in the same plane, so as to make the two interior angles on the same side of it, taken together, less than two right angles, these straight lines shall at length meet upon that side, if they be continually produced. Axiom twelfth, Elements of Euclid.*

Let EF be the straight line meeting AB, CD on the same plane, so that BLO, DOL are less than two right angles, the lines AB, CD will meet if continually produced.

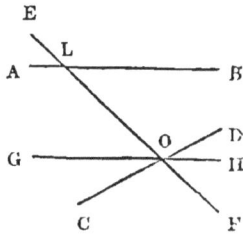

At the point O, draw GH (I. 18) parallel to AB, then BLO, HOL are equivalent to two right angles (I. 16), but DOL is less than HOL (I. ax. 9); therefore BLO, DOL are less than BLO, HOL, and less than two right angles. But GH, which forms with EF the angle HOL, is parallel (const.) with AB, which forms with EF the angle BLO, and (I. def. 11) GH and AB never meet each other, because (I. 15) they are equally distant from each other, and (I. 16, cor. 1) the interior angles HOL, BLO together equivalent to two right angles; therefore, then, CD, which forms with EF the angle DOL less than HOL, can not be parallel with AB (I. ax. 9); and not being parallel with AB, CD and AB can not preserve the condition of parallel lines (I. def. 11), and will meet. And the line CD making with EF the angle DOL less than HOL, on the side of EF, where the interior angles BLO, DOL are less than two right angles, the line OD must therefore be between GH and AB, and is consequently nearer to AB than GH is to AB (I. ax. 9); but CD making with EF the angle EOC greater than the angle GOE, which GH makes with EF, therefore CO must be without AB and GH, and consequently is farther from AB than GH is from AB; hence the straight line CD, which is made of CO and OD, has parts unequally distant from AB; therefore since CD approaches AB on the side of EF where OD is, CD must

3

meet AB on that side when continually produced, but OD
makes the angle DOL less than HOL, therefore CD will meet
AB on the side of EF where the angles BLO, DOL are less
than BLO, HOL. Wherefore, if a straight line, etc.

Cor. 1. Hence a straight line which intercepts one of two or
more parallel straight lines will intercept the others if continually produced; hence, also, two straight lines which intercept
each other are not both parallel to the same straight line.

Prop. XX.—Theor.—*If a side of any triangle be produced,*
(1) *the exterior angle is equivalent to the two interior and remote angles ; and* (2) *the three interior angles of every triangle
are together equivalent to two right angles.*

Let ABC be a triangle, and let one of its sides BC be produced to D ; (1) the exterior angle ACD is equivalent to the
two interior and remote angles CAB, ABC ; and (2) the three
interior angles, ABC, BCA, CAB, are together equivalent to
two right angles.

Through the point C draw (I. 18) CE parallel to the straight
line AB. Then, because AB is parallel to EC, and AC falls
upon them, the alternate angles BAC, ACE are (I. 16, part 1)
equal. Again: because AB is parallel to
EC, and BD falls upon them, the exterior
angle ECD is equivalent (I. 16, part 2)
to the interior and remote angle ABC ;
but the angle ACE has been shown to
be equal to the angle BAC ; there-
fore the whole exterior angle ACD is equivalent (I. ax. 2) to
the two interior and remote angles CAB, ABC. To these
equals add the angle ACB, and the angles ACD, ACB are
equivalent (I. ax. 2) to the three angles CBA, BAC, ACB ; but
the angles ACD, ACB are equivalent (I. 9) to two right angles ;
therefore, also, the angles CBA, BAC, ACB are equivalent to
two right angles. Wherefore, if a side, etc.

Another proof of this important proposition can be given by
producing the side AC through C to F. Now (I. 11), the angle
DCF is equal to the angle ACB ; EC being parallel (const.) to
BA, the exterior angle ECD is equal to the interior and remote
angle ABC (I. 16), and for the same reason the alternate angles

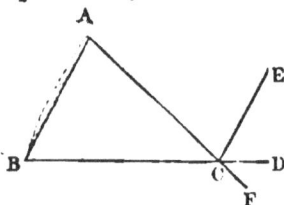

ECA and CAB are equal; hence we have the three angles of the triangle, ACB, CBA, and CAB, equal to the three angles DCF, DCE, and ECA, each to each; but (I. 9) the angles DCF, DCE, and ECA are equivalent to two right angles; therefore (I. ax. 1) the three angles ACB, CBA, and CAB of the triangle are likewise equivalent to two right angles. And when a parallelogram (I. def. 14) is formed by drawing (I. 18) parallels to BA and AC respectively, it can be shown by the sixteenth proposition that two adjacent angles of the parallelogram are equivalent to two right angles, and the four angles together equivalent to four right angles; since (I. 15, cor. 4) a diagonal bisects the parallelogram and forms two equal triangles, the angles are also equally divided, hence each triangle has its three angles equivalent to two right angles.

Cor. 1. All the interior angles of any rectilineal figure, together with four right angles, are equivalent to twice as many right angles as the figure has sides.

For any rectilineal figure can be divided into as many triangles as the figure has sides, by drawing straight lines from a point within the figure to each of its angles; and by the proposition, all the angles of these triangles are equivalent to twice as many right angles as there are triangles—that is, as there are sides of the figure; and the same angles are equivalent to the angles of the figure, together with the angles at the point which is the common vertex of the triangles—that is (I. 9, cor.), together with four right angles. Therefore all the angles of the figure, together with four right angles, are equivalent (I. ax. 1) to twice as many right angles as the figure has sides.

Scho. 1. Another proof of this corollary may be obtained by dividing the figure into triangles by lines drawn from any angles to all the remote angles. Then each of the two extreme triangles has two sides of the polygon for two of its sides, while each of the other triangles has only one side of the figure for one of its sides; and hence the number of triangles is less by two than the number of the sides of the figure. But the interior angles of the figure are evidently equivalent to all the interior angles of all the triangles—that is, to twice as many right angles as there are triangles, or twice as many right angles as the figure has sides, *less* the angles of two triangles—

that is, four right angles. Hence, in any equiangular figure, the number of the sides being known, the magnitude of each angle compared with a right angle can be determined. Thus, in a regular pentagon, the amount of all the angles being twice five right angles *less* four—that is, six right angles, each angle will be one fifth part of six right angles, or one right angle and one fifth. In a similar manner it would appear that in the regular hexagon, each angle is a sixth part of eight right angles, or a right angle and a third; that in the regular heptagon, each is a right angle and three sevenths; in the regular octagon, a right angle and a half, etc.

Cor. 2. All the exterior angles of any rectilineal figure are together equivalent to four right angles.

Because each interior angle ABC, and the adjacent exterior ABD, are together equivalent (I. 9) to two right angles, therefore all the interior, together with all the exterior angles of the figure, are equivalent to twice as many right angles as there are sides of the figure—that is, by the foregoing corollary, they are equivalent to all the interior angles of the figure, together with four right angles; therefore all the exterior angles are equivalent (I. ax. 3) to four right angles.

Scho. 2. It is to be observed, that if angles be taken in the ordinary meaning, as understood by Euclid, this corollary and the foregoing are not applicable when the figures have *re-entrant* angles—that is, such as open outward. The second corollary will hold, however, if the difference between each re-entrant angle and two right angles be taken from the sum of the other exterior angles; and the former will be applicable, if, instead of the angle which opens externally, the difference between it and four right angles be used. Both corollaries, indeed, will hold without change, if the re-entrant angle be regarded as internal and greater than two right angles; and if, to find the exterior angles, the interior be taken, in the algebraic sense, from two right angles, as in this case, the re-entering angles will give negative or subtractive results.

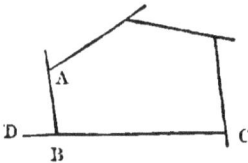

Cor. 3. If a triangle has a right angle, the remaining angles are together equivalent to a right angle; and if one angle of a triangle be equivalent to the other two, it is a right angle.

Cor. 4. The angles at the base of a right-angled isosceles triangle are each half a right angle.

Cor. 5. If two angles of one triangle be equal to two angles of another, their remaining angles are equal.

Cor. 6. Each angle of an equilateral triangle is one third of two right angles, or two thirds of one right angle.

Cor. 7. Hence, a right angle may be trisected by describing an equilateral triangle on one of the lines containing the right angle.

Scho. 3. By this principle also, in connection with the fifth proposition, we may trisect any angle, which is obtained by the successive bisection of a right angle, such as the half, the fourth, the eighth, of a right angle, and so on.

Cor. 8. Any two angles of a triangle are less than two right angles.

Cor. 9. Hence every triangle must have at least two acute angles.

PROP. XXI.—THEOR.—*If two sides of a triangle be unequal,* (1) *the greater side has the greater angle opposite to it ; and* (2) *conversely, if two angles of a triangle be unequal, the greater angle has the greater side opposite to it.*

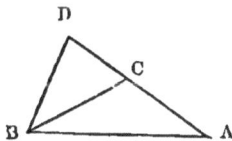

Let ABC be a triangle of which the side AB is greater than the side AC; the angle ACB opposite AB is greater than the angle ABC opposite AC.

Because AB is greater than AC, produce AC (I. post. 2), and with A as a center, and a radius AB, describe a circle (I. post. 3) intercepting AC produced in D, and join BD; the triangle ADB is isosceles (I. defs. 13 and 16); therefore the angle ADB is equal to the angle ABD (I. 1, cor. 1). But (I. 20) the exterior angle ACB is equivalent to the sum of the two remote interior angles CDB and DBC. And CDB is equal to DBA (I. 1, cor. 1), but ABC is less than ABD (I. ax. 9); therefore ACB is greater than CBA.

The proof can also be given by laying off on the greater side

AB, the less side AC; joining the vertex of the opposite angle with the point where AC terminates on AB; and the demonstration conducted similarly to the preceding. Again: by cutting off on AC a part equal to CB, bisect the angle BCA, and join the extremity of the part on AC equal to CB with the foot of the line bisecting the angle BCA; this proof is given by means of (I. 3 and 20).

Conversely: when the angle ACB is greater than the angle ABC, the side AB is greater than AC. At the vertex of BCA on AC make an angle ACD equal to the angle ABC (I. 13). Now, the angle ACD, the equal to the angle ABC, is subtended by AD, and the angle ACB is subtended by AB, but AB is greater than AD; hence the greater angle is subtended by the greater line; therefore, in the triangle ACB, the greater angle ACB is subtended by a greater side than the less angle ABC. And (I. def. 19) if the angles ACD and ABC be subtended by arcs, the arc subtending ACB is greater than the arc subtending ABC; but the side AB is the chord of the arc subtending ACB, and AC is the chord of the arc subtending ABC; therefore AB is greater than AC. Wherefore, if two sides of a triangle, etc.

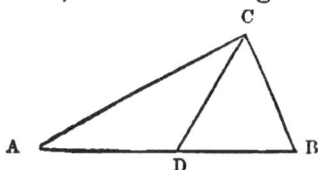

Cor. Hence any two sides of a triangle are together greater than the remaining side.

Scho. The truth of this corollary is so manifest, that it is given as a corollary to avoid increasing the number of axioms. Archimedes defined the straight line the shortest distance between two points; hence two straight lines connecting three points not in the same direction, are together greater than one straight line connecting any two of those points.

PROP. XXII.—THEOR.—*If two triangles have two sides of the one equal to two sides of the other, each to each, but the angles contained by those sides unequal, the base or remaining side of the one which has the greater angle is greater than the base or remaining side of the other.*

Let DEG and DEF be two triangles which have the sides DE common, the sides DG and DF equal, but the angle EDG greater than the angle EDF; the side EG is also greater than the side EF.

In the triangles DEF and DEG we have (hyp.) DE common, DG equal to DF, and the angle EDG greater than the angle EDF. Now, because DG and DF are equal, the angles DGF and DFG are equal (I. 1, cor. 1); but the angle DGF is greater than the angle EGF (I. ax. 9); therefore the angle DFG is greater than the angle EGF; and much more is the angle EFG greater than the angle EGF. Then (I. 21), EG opposite EFG is greater than EF opposite EGF. Wherefore, if two triangles have, etc.

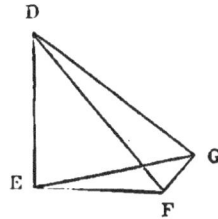

There are other cases of this proposition; if a line equal to DE, the less side, be drawn through D, making with DF, on the same side of it with DE, an angle equal to EDG, the extremity of that line might fall *on* FE produced, or *above*, or *below* it. Or the angle DFE could be *in* the triangle DEG, or *on the other* side of DE. And a very easy proof can be given by bisecting the angle FDG by a straight line cutting EG in a point, which call K, and joining DK and KF, for KG would be equal to KF (I. 3); adding EK, we would have EG equivalent to EK and KF; therefore (I. 21, cor.) EG greater than EF.

Cor. Hence, *conversely,* if two triangles have two sides of the one equal to two sides of the other, each to each, but their bases unequal, the angles contained by the respectively equal sides of those triangles are also unequal, the greater angle being in the triangle which has the greater base. For with a radius, DG, or its equal, DF, describe a circle (I. post. 3) from a center, D, and draw on the same side of DE with the angle EDF a line equal to EG from the other extremity of DE to the circumference, then in the triangles DEG and DEF we have (const.) DG equal to DF (I. def. 16). But (hyp.) the side EF is less than EG; hence the angle EDF is less than the angle EDG (I. ax. 9).

PROP. XXIII.—PROB.—*To describe a parallelogram upon a given straight line.*

Let DB be a given straight line; it is required to describe a

parallelogram upon it. From a point, D, on DB draw DF (I.
post. 1), then from the point F on
DF draw FL parallel to DB (I. 18);
and if through B, a point on DB, a
parallel to DF (I. 18), be drawn,
DBFG (I. def. 15) is a parallelogram.
If from D a perpendicular, DH (I. 7), be drawn, then the
angle HDB is a right angle (I. def. 10) ; and from H a parallel
to DB as HL be drawn (I. 18), and from B a parallel to HD
as LB be drawn (I. 18), then DBHL (I. def. 15) is a rectangle.
And if from D as a center, and a radius DB (I. post. 3), an arc
be described intercepting DH, or DH produced, and a rectangle
be described from that point where DH is intercepted, that
rectangle will be a square (I. def. 15).

Cor. 1. Hence (const. and I. 15, cor. 1) a square has all its
sides equal, and (I. 15, cor. 1) all its angles are right angles.

Cor. 2. Hence the squares described on equal straight lines
(I. 15, cor. 4) are equal.

Cor. 3. If two squares be equal, their sides are equal (I.
ax. 1).

Cor. 4. If AB and AD, two adjacent sides of a rectangle
BD, be divided into parts which are all equal, straight lines
drawn through the points of section, parallel to the sides, divide
the rectangle into squares which are all equal, and the number
of which is equal to the product of the
number of parts in AB, one of the sides,
multiplied by the number of parts in
AD, the other. For (const.) these
figures are all parallelograms; and (I.
def. 15 and const.) the sides being equal,
and the angles being (I. 16, part 2) equal to A, and therefore right angles, hence (I. 15, cor. 4) they are all squares. Of
these squares, also, there are evidently as many *columns* as there
are parts in AB ; while in each column there are as many squares
as there are parts in AD. The number of such squares contained in a figure is called, in the language of mensuration, the
area of that figure.

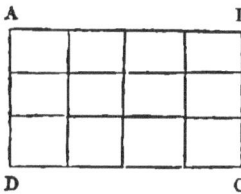

Cor. 5. Hence, since any parallelogram is equivalent (I. 15,
cor. 5) to a rectangle on the same base and between the same

parallels, it follows that *the area of any parallelogram is equivalent to the product of its base and its perpendicular height ; and the area of a square is computed by multiplying a side by itself.*

Cor. 6. Hence, also (I. 15, cor. 4), *the area of a triangle is computed by multiplying any of its sides by the perpendicular drawn to that side from the opposite angle, and taking half the product; and the area of a trapezium is found by multiplying either diagonal by the sum of the perpendiculars drawn to it from the angles which it subtends, and taking half the product.* When, in consequence of one of the angles being re-entrant, the perpendiculars lie on the same side of the diagonal, the *difference* of the perpendiculars must evidently be used instead of their sum.

Cor. 7. Every polygon may be divided into triangles or trapeziums by drawing diagonals; and therefore the area of any polygon whatever can be computed by finding the areas of those component figures by the last corollary, and adding them together.

Scho. This corollary and the two foregoing contain the elementary principles of the mensuration of rectilineal figures, and they form a connection between arithmetic or algebra and geometry. They also explain the origin of the expressions, "the square of a number," "the rectangle of two numbers," and "the product of two lines."

PROP. XXIV.—THEOR.—*If parallelograms be described on two sides of any triangle, and their sides which are parallel to the sides of the triangle be produced until they meet, the sum of the parallelograms will be equivalent to the parallelogram described on the base of the triangle having its adjacent sides to the base parallel to the straight line joining the vertex of the triangle with the point of intersection of the sides of the other parallelograms produced, and terminated by the latter sides or those sides produced.*

Let BAC be a triangle; the parallelograms MBAD and CEFA, described on the two sides BA and CA, respectively, are together equivalent to the parallelogram BCHK, described on the base BC; the parallelograms MBAD, CEFA, and BCHK being described agreeably to the proposition.

Describe on BA and CA (L 23) the parallelograms MBAD and CEFA; and produce the sides MD and EF until they meet in G; draw GA, and produce it to L on the base BC (I. post. 2); describe the parallelogram BCHK (I. 23) on BC. Then, since (const. and I. 15, cor. 1) BH and CK are parallel and equal to AG, they are parallel and equal to one another (I. ax. 1); also (I. 15, cor 3) HK is parallel and equal to BC; hence (I. def. 15) BCHK is a parallelogram, and BLHW and LCWK are also parallelograms. Now, the parallelograms BLHW and HBAG (I. 15, cor. 5) are equal, and for similar reason the parallelograms HBAG and MBDA; hence (I. ax. 1) MBDA is equal to BLHW. In similar manner, it can be shown that CEFA is equivalent to LCKW; therefore the whole parallelogram BCHK (I. axs. 2 and 10) is equivalent to the sum of the two parallelograms MBAD and CEFA. Wherefore, if on any two sides of a triangle, etc.

Cor. 1. A particular case of this proposition is, *when the triangle is right-angled, then the squares described on the legs— that is, the sides containing the right angle, are together equivalent to the square on the hypothenuse—that is, the side opposite the right angle.*

Let ABC be a right-angled triangle, having the right angle BAC; the square described on the hypothenuse BC is equivalent to the sum of the squares on BA and AC. On BC, BA, and AC (I. 23) describe the squares BCHK, BADE, and ACFG; through A draw AL parallel to BH (I. 18), and draw AH and DC (I. post. 1).

Then, because the angles BAC and BAE are both right angles (I. def. 10), the two straight lines CA and AE are in the same straight line (I. 10). For like reason, BA and AF are in the same straight line. Again: because the angle HBC is equal to the angle DBA (I. ax. 1), if the angle ABC be added to each, we have (I. ax. 2) HBA equal to DBC; and because AB is equal to DB (const.), and BH equal to BC, therefore (I. 3, case 1) the triangle ABH is equal to the triangle DBC. But (I. 15, cor. 4) the parallelogram BPHL is double the triangle ABH; for like reason the square BADE is double the triangle DBC; hence (I. ax. 6) BPHL is equivalent to BADE. In like manner, PCLK can be shown equivalent to ACFE. Now (I. ax. 10), BCHK is equivalent to BPHL and PCLK together; hence (I. ax. 1) BCHK is equivalent to BADE and ACFG together. Wherefore, if the triangle is right-angled, etc.

OTHERWISE,

Let the squares on AB and BC fall on the same side of BC. Describe the square BAED on the side BA (I. 23), and the square BCMN on the side BC (I. 23), and produce AE to F (I. post. 2); then through D draw PF parallel to BC (I. 18).

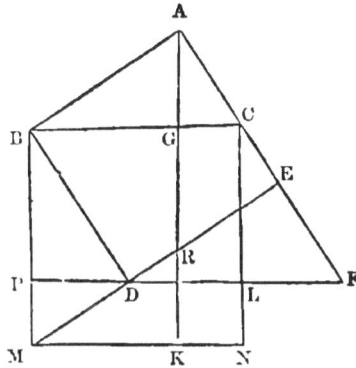

Because AF is parallel to BD (I. def. 14, and 15, cor. 1), BC is equal to DF, and BA is equal to DE, and the angles BAC and DEF are both right angles (I. def. 10), and equal (I. ax. 11); therefore the triangle BAC is equal to the triangle DEF (I. 3); and because BCED is common to the square BAED and the parallelogram (I. def. 14) BCDF, and the triangles BAC and DEF equal, the square BAED is equivalent to the parallelogram BCDF. And PF being parallel to BC (const.), the parallelograms (I. def. 14) BCDF and BCPL have a common base, BC, and equal altitudes; hence (I. 15, cor. 5) they are equivalent, and (I. ax. 1) the square BAED is equivalent to the parallelogram BCPL. From A draw (I. 18) AK parallel to BM, and produce DE (I. post. 2) to BM; then AK and BM being parallel (const.), ED and

BA, being opposite sides of the same square, are also parallel
(I. def. 15); hence MR is equal to BA, and equal also to DE
(I. ax. 1). But BM is equal to BC (const.); therefore the par-
allelogram BAMR is equal to the parallelogram BCDF; and
BAMR having the same base and equal altitude with the par-
allelogram BGMK, is equivalent to it (I. 15, cor. 5); hence
BGMK is equivalent to BCDF, equivalent to BCPL, and
equivalent to the square BADE (I. ax. 1).

Or, *the square described upon BA is equivalent to the rectangle of the hypothenuse BC and the part BG of the hypothenuse nearest to BA intercepted by the perpendicular drawn
from the vertex of the right angle to the hypothenuse. In a
similar manner, it can be shown that the square described on
AC is equivalent to the rectangle of the hypothenuse BC, and
the remaining part GC of the hypothenuse intercepted by the
same perpendicular.* But the two rectangles are equivalent to
the square of the hypothenuse (I. ax. 10); hence *the two squares
described on the sides AB and AC* (I. ax. 1) *are equivalent to
the square described on the hypothenuse.*

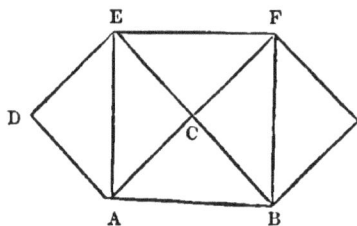

And if we make the triangle
an isosceles right-angled triangle as ABC, the square described on AB will contain
four equal triangles, ACB, BCF,
FCE, and ECA, while each of
the squares described on AC
and CB will contain two such triangles, and both together will
be equivalent to the four equal triangles, or equivalent to the
square on AB. The demonstration is very simple, and it would
be well for the pupil to undertake it.

Hence, *conversely,* if the square described upon one side of a
triangle be equivalent to the sum of the square described upon
the two other sides of the triangle, the angle contained by those
two sides is a right angle; and when those two sides form two
equal squares, the triangle is a right-angled isosceles triangle.

Scho. 1. The proof of the corollary can be shown, also, either
by describing the square of the hypothenuse on the other side
of BC; and the other squares sometimes on one side and sometimes on the other; and since drawing a perpendicular from the

vertex of the right angle to the hypothenuse makes two *right angles*, so a line can be drawn from the same vertex to the point of bisection of the hypothenuse and make two *supplemental angles* and two equivalent triangles, and the demonstration conducted by *supplemental* angles (I. 15, cor. 7) instead of right angles. Proportion also gives neat and easy solutions to this corollary. (See V. 8, *scho.*)

Cor. 2. If two right-angled triangles have their hypothenuses equal, and a side similarly situated in each also equal, the two triangles are equal by the third proposition of this book; and, *conversely*, if the *legs* of a right-angled triangle be equal to the *legs* of another right-angled triangle, each to each, their hypothenuses can be in a similar manner shown equal.

Cor. 3. Hence, also, we can find a square equivalent to the sum of more than two squares; thus, let AB be the side of one square, and AC, perpendicular to it, the side of another square; join CB; the square on CB (I. 24, cor. 1) is equivalent to the sum of the squares on CA and AB. In like manner, if CD be drawn perpendicular to CB, and DB be drawn, the square on DB is equivalent to the squares on DC, CA, and AB, and by drawing a perpendicular to DB, a square can be found equivalent to the sum of four squares; hence a square can be found equivalent to the sum of any number of squares.

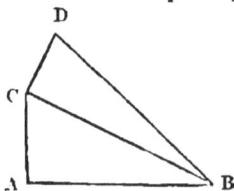

Cor. 4. Since $(CB)^2 = (AB)^2 + (CA)^2$, we have $(AB)^2 = (CB)^2 — (CA)^2$; hence a square can be found equivalent to the difference of two squares.

Cor. 5. If a perpendicular be drawn from the vertex of the angle A, in the triangle BAC (diagram to cor. 1), to P on the hypothenuse BC, cutting BC into two segments, BP and PC, the difference of the squares on the sides AB and AC is equivalent to the difference of the squares on the segments BP and PC. For the square on AB is equivalent to the squares on BP and PA, and the square on AC is equivalent to the squares on PC and PA; therefore $(AC)^2 — (AB)^2 = (PC)^2 — (BP)^2$.

Cor. 6. Hence *the squares on two sides of a triangle are together equivalent to twice the square of half the remaining side, and twice the square of the straight line from its point of bi-*

section to the opposite angle. Suppose P in the triangle to be the point of bisection of the side BC; then, when AP is perpendicular to BC, we have $(AB)^2 + (AC)^2 = (BP)^2 + (AP)^2 + (PC)^2 + (AP)^2 = 2 (BP)^2 + 2 (AP)^2$. And when AP is not perpendicular to BC, the equivalence of the squares on two sides to the same will be shown in the next book.

Scho. 2. In proof of cor. 5, the obvious principle is employed, that the difference of two magnitudes is the same as the difference obtained after adding to each the same third magnitude. Thus the difference of the squares on BP and PC is the same as the difference between the sum of the squares BP and PA and of PC and PA.

Prop. XXV.—Theor.—*The side and diagonal of a square are incommensurable to one another—that is, there is no line which is a measure of both.*

Let ABCD be a square, and BD one of its diagonals; AB, BD are incommensurable.

Cut off DE equal to DA, and join AE. Then, since (I. 1, cor. 1) the angle DEA is equal to the acute angle DAE, AEB

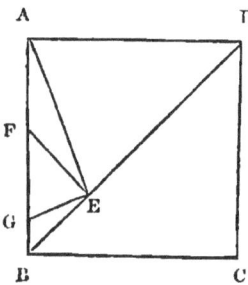

is obtuse, and therefore (I. 22, cor.) in the triangle ABE, BE is less than AB, or than AD; wherefore AD is not a measure of BD. Draw EF perpendicular to BD. Then the angles FAE, FEA, being the complements of the equal angles DAE, DEA, are equal, and therefore AF, FE are equal. But (I. 20, cor. 4) ABD is half a right angle; as is also BFE, since BEF is a right angle; wherefore BE is equal to FE, and therefore to AF. From FB, which is evidently the diagonal of a square of which FE or EB is the side, cut off FG equal to FE, and join GE. Then it would be shown, as before, that BG is less than BE; and therefore BE, the difference between the side and diagonal of the square AC, is contained twice in the side AB, with the remainder GB, which is itself the difference between the side FE or EB, and the diagonal FB of another square. By repeating the process, we should find, in exactly the same manner, that BG would be contained twice in BE, with a remainder, which would be the difference between the side and diagonal of a square described on BG; and it is evident that a like process might be repeated continually, as no excess of a diagonal above a side would be contained in the side without remainder; and as this process has no termination, there is no line, however small, which will be contained without remainder in both AB and BD; they are, therefore, incommensurable.

Scho. This proposition can be illustrated by numbers. Let 10 be the side of the square; then (I. 24, cor. 1) the diagonal will be expressed by the square root of 200, or 14·142 +; there being no common multiple of 10 and 14·142 +, these numbers are incommensurable with each other. Or, when any two lines are taken which by division and subdivision no common measure can be found which can be contained in each without a remainder, the two lines are said to be *incommensurable* with each other, and such lines are the *side* and *diagonal* of a *square;* and any two magnitudes whatever which have no common unit of measure are incommensurable with one another.

END OF BOOK FIRST.

BOOK SECOND.*

DEFINITIONS.

1. A *rectangle* is said to be *contained* by the two straight lines which are about any of the right angles. For the sake of brevity, the rectangle contained by AB and CD is often expressed simply by AB. CD, a point being placed between the letters denoting the sides of the rectangle; and the square of a line AB is often written simply AB².

2. A *gnomon* is the part of a parallelogram which remains when either of the parallelograms about one of the diagonals is taken away.

PROPOSITIONS.

PROP. I.—THEOR.—*If there be two straight lines, one of which is divided into any number of parts, the rectangle contained by the two lines is equivalent to the rectangles contained by the undivided line, and the several parts of the divided line.*

Let A and BC be two straight lines; and let BC be divided into any parts in the points D, E; the rectangle contained by A and BC is equivalent to the rectangles contained by A and BD, A and DE, and A and EC.

From B draw (I. 7) BG perpendicular to BC, and make it equal to A; through G draw (I. 18) GH parallel to BC; and through D, E, C draw DK, EL, CH parallel to BG. Then BH, BK, DL, and EH are evidently rectangles; and BH is equivalent (I. ax. 10) to BK, DL, EH. But BH is contained by A and BC, for (II. def. 1) it is contained by GB and BC, and GB is

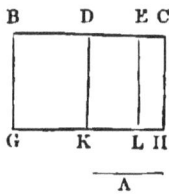

---

* The second and third books are arranged very similarly to those books in the edition of Euclid by Professor Thomson of the University of Glasgow, Scotland.

equal (const.) to A; and BK is contained by A and BD, for it is contained by GB and BD, of which GB is equal to A. Also DL is contained by A and DE, because DK, that is (I. 15, cor. 4) BG, is equal to A; and in like manner it is shown that EH is contained by A and EC. Therefore the rectangle contained by A and BC is equivalent to the several rectangles contained by A and BD, by A and DE, and by A and EC. Wherefore, if there be two straight lines, etc.

PROP. II.—THEOR.—*If a straight line be divided into any two parts, the rectangles contained by the whole and each of the parts are together equivalent to the square of the whole line.*

Let the straight line AB be divided into any two parts in the point C; the rectangles AB.AC and AB.BC are equivalent to the square of AB.

Upon AB describe (I. 23) the square AE, and through C draw (I. 18) CF parallel to AD or BE. Then AE is equal (I. ax. 10) to the rectangles AF and CE. But AE is the square of AB, and AF is the rectangle contained by BA, AC; for (II. def. 1) it is contained by DA, AC, of which DA is equal to AB; and CE is contained by AB, BC, for BE is equivalent to AB; therefore the rectangles under AB, AC, and AB, BC are equivalent to the square of AB. If, therefore, etc.

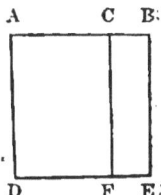

Scho. This proposition may also be demonstrated in the following manner:

Take a straight line D equal to AB. Then (II. 1) the rectangles AC.D and BC.D are together equivalent to AB.D. But since D is equal to AB, the rectangle AB.D is equivalent (I. def. 15) to the square of AB, and the rectangles AC.D and BC.D are respectively equivalent to AC.AB and BC.AB; wherefore the rectangles AC.AB and BC.AB are together equivalent to the square of AB.

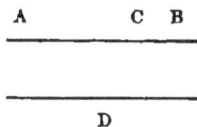

In a manner similar to this, several of the following propositions may be demonstrated. Such proofs, though perhaps not so easily understood at first by the learner, are shorter than

4

those given by Euclid; and they have the advantage of being
derived from those preceding them, instead of being established by continual appeals to original principles.

Prop. III.—Theor.—*If a straight line be divided into any
two parts, the rectangle contained by the whole and one of. the
parts is equivalent to the square of that part, together with the
rectangle contained by the two parts.*

Let the straight line AB be divided into two parts in the
point C; the rectangle AB.BC is equivalent to the square of
BC, together with the rectangle AC.CB

Upon BC describe (I. 23) the square CE; produce ED to F;
and through A draw (I. 18) AF parallel
to CD or BE. Then the rectangle AE
is equivalent (I. ax. 10) to the rectangles
CE, AD. But AE is the rectangle contained by AB, BC, for it is contained
by AB, BE, of which BE is equal to
BC; and AD is contained by AC, CB,
for CD is equal to CB; also DB is the square of BC. Therefore the rectangle AB.BC is equivalent to the square of BC,
together with the rectangle AC.CB. If, therefore, etc.

Scho. Otherwise: Take a line D equal to CB. Then (II. 1)
the rectangle AB.D is equivalent to the
rectangles BC.D and AC.D; that is (const.
and I. def. 15), the rectangle AB.BC is
equivalent to the square of BC together with
the rectangle AC.CB.

Prop. IV.—Theor.—*If a straight line be divided into
any two parts, the square of the whole line is equivalent to
the squares of the two parts, together with twice their rectangle.*

Let the straight line AB be divided into any two parts in C;
the square of AB is equivalent to the squares of AC and CB,
together with twice the rectangle under AC and CB.

On AB describe (I. 23) the square of AE, and join BD;
through C draw (I. 18) CGF parallel to AD or BE; and
through G draw HK parallel to AB or DE. Then, because

CF is parallel to AD, and BD falls upon them, the exterior angle CGB is equivalent (I. 16) to the interior and remote angle ADB; but ADB is equal (I. 1) to ABD, because BA and AD are equal, being sides of a square; wherefore (I. ax. 1) the angle CGB is equal to GBC; and therefore the side BC is equal (I. 1, cor.) to the side CG. But (const.) the figure CK is a parallelogram; and since CBK is a right angle, and BC equal to CG, CK (I. def. 15) is a square, and it is upon the side CB. For the like reason HF also is a square, and it is upon the side HG, which is equal (I. 15, cor. 1) to AC; therefore HF, CK are the squares of AC, CB. And because (I. 15, cor. 8) the complements AG, GE are equivalent, and that AG is the rectangle contained by AC, CB, for CG has been proved to be equal to CB; therefore GE is also equivalent to the rectangle AC.CB; wherefore AG, GE are equivalent to twice the rectangle AC.CB. The four figures, therefore, HF, CK, AG, GE are equivalent to the squares of AC, CB, and twice the rectangle AC.CB. But HF, CK, AG, GE make up the whole figure AE, which is the square of AB; therefore the square of AB is equivalent to the squares of AC and CB, and twice the rectangle AC.CB. Wherefore, etc.

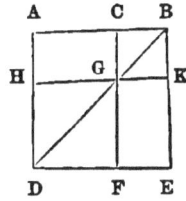

Otherwise: $AB^2 = AB.AC + AB.CB$ (II. 2). But (II. 3) $AB.AC = AC^2 + AC.CB$, and $AB.CB = CB^2 + AC.BC$. Hence (I. ax. 2) $AB.AC + AB.BC$, or $AB^2 = AC^2 + CB^2 + 2AC.CB$.

Cor. 1. It follows from this demonstration, that the parallelograms about the diagonal of a square are likewise squares.

Cor. 2. Hence the square of a straight line · is equivalent to four times the square of its half; for the straight line being bisected, the rectangle of the parts is equivalent to the square of one of them.

PROP. V.—THEOR.—*If a straight line be divided into two equal parts, and also into two unequal parts; the rectangle contained by the unequal parts, together with the square of the line between the points of section, is equivalent to the square of half the line.*

Let the straight line AB be bisected in C, and divided unequally in D; the rectangle AD.DB, together with the square of CD, is equivalent to the square of CB.

On CB describe (I. 23) the square CF, join BE, and through D draw (I. 18) DHG parallel to CE or BF; also through H draw KLM parallel to CB or EF; and through A draw AK parallel to CL or BM. Then (I. 15, cor. 5) AL and CM are equal, because AC is equal to CB; and (I. 15, cor. 8) the complements CH and HF are equivalent. Therefore (I. ax. 2) AL and CH together are equal to CM and HF together; that is, AH is equivalent to the gnomon CMG. To each of these add LG, and (I. ax. 2) the gnomon CMG, together with LG, is equivalent to AH together with LG. But the gnomon CMG, and LG make up the figure CEFB, which is the square of CB; also AH is the rectangle under AD and DB, because DB is equal (II. 4, cor. 1) to DH; and LG is the square of CD. Therefore the rectangle AD.DB and the square of CD are equivalent to the square of CB. Wherefore, if a straight line, etc.

Otherwise: Since, as is easily seen from the proof in the above, DF is equal to AL, take these separately from the entire figure, and there remain AH and LG equivalent to the square CF, as before. The proof may also be as follows:

AD.DB=CD.DB+AC.DB (II. 1) or AD.DB=CD.DB+ CB.DB, because CB=AC. Hence (II. 3) AD.DB=CD.DB+ CD.DB+DB²=2CD.DB+DB². To each of these add CD²; then AD.DB+CD²=2CD.DB+DB²+CD², or (II. 4) AD.DB +CD²=CB².

Cor. 1. Hence the difference of the squares of CB and CD is equivalent to the rectangle under AD and DB. But since AC is equal to CB, AD is equivalent to the sum of CB and CD, and DB is the difference of these lines. Hence *the difference of the squares of two straight lines is equivalent to the rectangle under their sum and difference.*

Cor. 2. Since the square of CB, or, which is the same, the rectangle AC.CB, is greater than the rectangle AD.DB by the square of CD, it follows, that to divide a straight line into two parts, the rectangle of which may be the greatest possible, or, as is termed, a *maximum*, the line is to be bisected.

Cor. 3. Hence also the sum of the squares of the two parts into which a straight line is divided, is the least possible, or is, as it is termed, a *minimum*, when the line is bisected. For (II. 4) the square of the line is equivalent to the squares of the parts and twice their rectangle ; and therefore the greater the rectangle is, the less are the squares of the parts; but, by the foregoing corollary, the rectangle is a maximum when the line is bisected.

Cor. 4. Since (I. 24, cor. 5) the difference of the squares of the sides of a triangle is equivalent to the difference of the squares of the segments of the base, it follows, from the first corollary above, that the rectangle under the sum and difference of the sides of a triangle is equivalent to the rectangle under the sum and difference of the segments, intercepted between the extremities of the base and the point in which the perpendicular cuts the base, or the base produced.

Cor. 5. Hence, also, if a straight line be drawn from the vertex of an isosceles triangle to any point in the base, or its continuation, the difference of the squares of that line and either of the equal sides is equivalent to the rectangle under the segments intercepted between the extremities of the base and the point.

Cor. 6. Since (I. 24, cor. 4) the square of one of the legs of a right-angled triangle is equivalent to the difference of the squares of the hypothenuse and the other leg, it follows (II. 5, cor. 1) that the square of one leg of a right-angled triangle is equivalent to the rectangle under the sum and difference of the hypothenuse and the other.

Scho. And a parallelogram can be constructed equivalent to a given triangle, or any given rectilinear figure having an angle equal to a given angle, or *applied* to a given straight line —that is, having that straight line for one of its sides, when the parallelogram shall be equivalent to a given triangle or given rectilinear figure, and have one of its angles equal to a given angle, by applying a parallelogram equivalent to the given triangle with an equal angle, to a given straight line, and then constructing an equal triangle to the given triangle (I. 15 and 15, cor. 4).

PROP. VI.—THEOR.—*If a straight line be bisected, and be produced to any point, the rectangle contained by the whole line thus produced, and the part of it produced, together with the square of half the line bisected, is equivalent to the square of the straight line which is made up of the half and the part produced.*

Let the straight line AB be bisected in C, and produced to D; the rectangle AD.DB, and the square of CB, are equivalent to the square of CD.

Upon CD describe (I. 23) the square CF, and join DE; through B draw (I. 18) BHG parallel to CE or DF; through H draw KLM parallel to AD or EF, and through A draw AK parallel to CL or DM. Then because AC is equal to CB, the rectangle AL is equal (I. 15, cor. 5) to CH; but (I. 15, cor. 8) the complements CH, HF are equivalent; therefore, also, AL is equal to HF. To each of these add CM and LG; therefore AM and LG are equivalent to the whole square CEFD. But AM is the rectangle under AD and DB, because (II. 4, cor. 1) DB is equal to DM; also, LG is the square of CB, and CEFD the square of CD. Therefore the rectangle AD.DB and the square of CB are equivalent to the square of CD. Wherefore, if a straight line, etc.

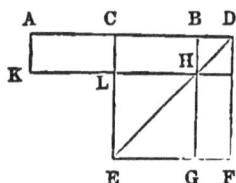

Otherwise: Produce CA to N, and make CN equal to CD. To these add the equals CB and CA; therefore NB is equal (I. ax. 2) to AD. But (II. 5) the rectangle NB.BD, or AD.BD, together with the square of CB, is equivalent to the square of CD; which was to be proved.

PROP. VII.—THEOR.—*If a straight line be divided into any two parts, the squares of the whole line and of one of the parts are equivalent to twice the rectangle contained by the whole and that part, together with the square of the other part.*

Let the straight line AB be divided into any two parts in the point C; the squares of AB, BC are equivalent to twice the rectangle AB.BC, together with the square of AC.

Upon AB describe (I. 23) the square AE, and construct the figure as in the preceding propositions. Then, because (I. 15, cor. 8) the complements CH, FK are equivalent, add to each of them CK; the whole AK is therefore equal to the whole CE; therefore AK, CE are together double of AK. But AK, CE are the gnomon AKF, together with the square CK; therefore the gnomon AKF and the square CK are double of AK, or double of the rectangle AB.BC, because BC is equal (II. 4, cor. 1) to BK. To each of these equals add HF, which is equal (II. 4, cor. 1) to the square of AC; therefore the gnomon AKF, and the squares CK, HF are equivalent to twice the rectangle AB.BC and the square of AC. But the gnomon AKF, and the squares CK, HF make up the whole figure AE, together with CK; and these are the squares of AB and BC; therefore the squares of AB and BC are equivalent to twice the rectangle AB.BC, together with the square of AC; wherefore, if a straight line, etc.

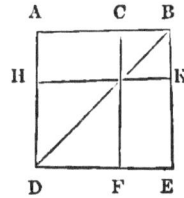

Otherwise: Because (II. 4) $AB^2 = AC^2 + BC^2 + 2AC.BC$, add BC^2 to both; then $AB^2 + BC^2 = AC^2 + 2BC^2 + 2AC.BC$. But (II. 3) $BC^2 + AC.BC = AB.BC$, and therefore $2BC^2 + 2AC.BC = 2AB.BC$; wherefore $AB^2 + BC^2 = AC^2 + 2AB.BC$.

Cor. 1. Since AC is the difference of AB and BC, it follows that the square of the difference of two straight lines is equivalent to the sum of their squares, wanting twice their rectangle.

Cor. 2. Since (II. 4) the square of the sum of two lines exceeds the sum of their squares by twice their rectangle, **and** since, by the foregoing corollary, the square of their difference is less than the sum of their squares by twice their rectangle, it follows that the square of the sum of two lines is equivalent **to** the square of their difference, together with **four times their** rectangle.

PROP. VIII.—THEOR.—*If a straight line be divided into two equal, and also into two unequal parts, the squares of the unequal parts are together double of the square of half the line, and of the square of the line between the points of section.*

Let the straight line AB be divided equally in C, and unequally in D; the squares of AD, DB are together double of the squares of AC, CD.

From C draw (I. 7) CE perpendicular to AB, and make it equal to AC or CB, and join EA, EB; through D draw (I. 18) DF parallel to CE, and through F draw FG parallel to AB; and join AF. Then because (const.) the triangles ACE, BCE are right-angled and isosceles, the angles CAE, AEC, CEB, EBC are (I. 20, cor. 4) each half a right angle. So also (I. 16, part 2) are EFG, BFD, because FG is parallel to AB, and FD to EC; and for the same reason EGF, ADF are right angles. The angle AEB is also a right angle, its parts being each half a right angle. Hence (I. 1) EG is equal to GF or CD, and FD to DB. Again (I. 24, cor. 1): the square of AE is equivalent to the squares of AC, CE, or to twice the square of AC, since AC and CE are equal. In like manner, the square of EF is equivalent to twice the square of GF or CD. Now (I. 24, cor. 1), the squares of AD and DF, or of AD and DB, are equivalent to the square of AF; and the squares of AE, EF, that is, twice the square of AC and twice the square of CD, are also equivalent to the square of AF; therefore (I. ax. 1) the squares of AD, DB are equivalent to twice the square of AC and twice the square of CD. If, therefore, a straight line, etc.

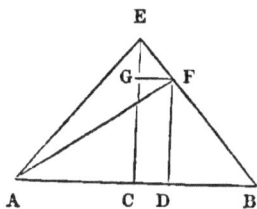

Otherwise: $DB^2 + 2BC.CD = BC^2 + CD^2$ (II. 7), or $DB^2 + 2AC.CD = AC^2 + CD^2$; also (II. 4) $AD^2 = AC^2 + CD^2 + 2AC.CD$. Add these equals together, and from the sums take $2AC.CD$; then $AD^2 + DB^2 = 2AC^2 + 2CD^2$.

PROP. IX.—THEOR.—*If a straight line be bisected, and produced to any point, the squares of the whole line thus produced, and of the part of it produced, are together double of the square of half the line bisected, and of the square of the line made up of the half and the part produced.*

Let the straight line AB be bisected in C, and produced to D; the squares of AD, DB are double of the squares of AC, CD.

From C draw (I. 7) CE perpendicular to AB; and make it

equal to AC or CB; join AE, EB, and through E and D draw
(I. 18) EF parallel to AB, and
DF parallel to CE. Then because the straight line EF
meets the parallels EC, FD,
the angles CEF, EFD are
equivalent (I. 9) to two right
angles; and therefore the angles BEF, EFD are less than
two right angles; therefore (I. 19) EB, FD will meet, if produced toward B, D; let them meet in G, and join AG. Then
it would be proved, as in the last proposition, that the angles
CAE, AEC, CEB, EBC are each half a right angle, and AEB
a right angle. BDG is also a right angle, being equal (I. 16)
to ECB, since (const.) EC, FG are parallel; DGB, DBG are
each half a right angle, being equal (I. 16 and I. 11) to CEB,
CBE, each to each; and FEG is half a right angle, being (I.
16) equal to CBE. It would also be proved, as in the last
proposition, that the square of AE is twice the square of AC,
and the square of EG twice the square of EF or CD. Now
(I. 24, cor. 1), the squares of AD, DG, or of AD, DB, are equivalent to the square of AG; and the squares of AE, EG, or twice
the square of AC and twice the square of CD, are also equivalent to the square of AG. Therefore (I. ax. 1) the squares of
AD, DB are equivalent to twice the square of AC and twice
the square of CD. If, therefore, a straight line, etc.

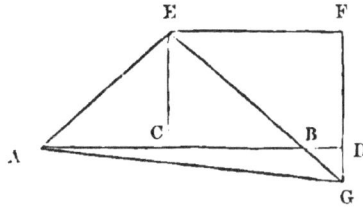

Otherwise: Produce CA, making CH equal to CD. To
these add CB, CA; therefore HB, AD are
equal. Then (II. 8) $HB^2 + BD^2$, or $AD^2 +$
$BD^2 = 2CD^2 + 2AC^2$.

Scho. The nine foregoing propositions may all be proved
very easily by means of algebra, in connection with the principles of mensuration, already established in the corollaries to
the 23d proposition of the first book. Thus, to prove the fourth
proposition, let AC = a, CB = b, and, consequently, AB = $a + b$.
Now, the area of the square described on AB will be found (I.
23, cor. 4) by multiplying $a + b$ by itself. This product is
found, by performing the actual operation, to be $a^2 + 2ab + b^2$;
an expression, the first and third parts of which are, by the

same corollary, the areas of the squares of AC and CB, and the second is twice the rectangle of those lines.

In like manner, to prove the eighth, adopting the same notation, we have the line which is made up of the whole and CB $=a+2b;$ and, multiplying this by itself, we get for the area of the square of that line, $a^2+4ab+4b^2$, or $a^2+4(a+b)b$, the first part of which is the area of the square of AC, and the second four times the area of the rectangle under AB and CB.

It will be a useful exercise for the student to prove the other propositions in a similar manner. He will also find it easy to investigate various other relations of lines and their parts by means of algebra.

All the properties delivered in these propositions hold also respecting numbers, if products be substituted for rectangles. Thus, 7 being equal to the sum of 5 and 2, the square, or second power of 7, is equal to the squares of 5 and 2 and twice their product; that is, $49=25+4+20$.

PROP. X.—PROB.—*To divide a given straight line into two parts, so that the rectangle contained by the whole and one of the parts may be equivalent to the square of the other part.*

Let AB be the given straight line; it is required to divide it into two parts, so that the rectangle under the whole and one of the parts may be equivalent to the square of the other.

Upon AB describe (I. 23) the square AD; bisect (I. 6) AC in E, and join E with B, the remote extremity of AB; produce CA to F, making EF equal to EB, and cut off AH equal to AF; AB is divided in H, so that the rectangle AB.BH is equivalent to the square of AH.

Complete the parallelogram AG, and produce GH to K. Then, since BAC is a right angle, FAH is also (I. 9) a right angle; and (I. def. 15) AG is a square, AF, AH being equal by construction. Because the straight line AC is bisected in E, and produced to F, the rectangle CF.FA and the square of AE are together equivalent (II. 6) to the square of EF or of EB, since (const.) EB, EF are equal. But the squares of BA, AE are equivalent (I. 24, cor. 1) to the square of EB, because the angle EAB is a right angle; therefore the rectangle CF.FA and the square of AE are equivalent (I. ax. 1) to the squares of

BA, AE. Take away the square of AE, which is common to both; therefore the remaining rectangle CF.FA is equivalent (I. ax. 3) to the square of AB. But the figure FK is the rectangle contained by CF, FA, for AF is equal to FG; and AD is the square of AB; therefore FK is equal to AD. Take away the common part AK, and (I. ax. 3) the remainders FH and HD are equivalent. But HD is the rectangle AB.BH, for AB is equal to BD; and FH is the square of AH. Therefore the rectangle AB.BH is equivalent to the square of AH; wherefore the straight line AB is divided in H, so that the rectangle AB.BH is equal to the square of AH; which was to be done.

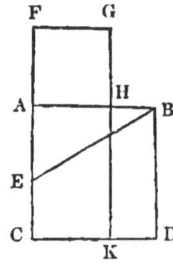

Scho. 1. In the practical construction in this proposition, and in the 2d cor. to 9 of the fifth book, which is virtually the same, it is sufficient to draw AE perpendicular to AB, making it equal to the half of AB, and producing it through A; then, to make EF equal to the distance from E to B, and AH equal to AF. It is plain that BD might be bisected instead of AC, and that in this way another point of section would be obtained.

While the enunciation in the text serves for ordinary purposes, it is too limited in a geometrical sense, as it comprehends only one case, excluding another. The following includes both:

In a given straight line, or its continuation, to find a point, such that the rectangle contained by the given line, and the segment between one of its extremities and the required point, may be equal to the square of the segment between its other extremity and the same point.

The point in the continuation of BA will be found by cutting off a line on EC and its continuation, equal to EB, and describing on the line composed of that line and AE a square lying on the opposite side of AC from AD; as the angular point of that square in the continuation of BA is the point required. The proof is the same as that given above, except that a rectangle corresponding to AK is to be added instead of being subtracted.

Scho. 2. The line CF is equivalent to BA and AH; and since it has been shown that the rectangle CF.FA is equivalent to the square of BA or CA, it follows that if any straight line AB (see the next diagram) be divided according to this proposition in C, AC being the greater part, and if AD be made equal to AB, DC is similarly divided in A. So also if DE be made

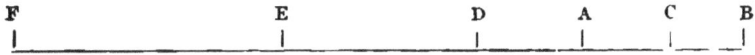

F   E   D   A   C   B
|_____|_____|_____|_____|___ _|

equal to DC, and EF to EA, EA is divided similarly in D, and FD in E; and the like additions may be continued as far as we please.

Conversely, if any straight line FD be divided according to this proposition in E, and if EA be made equal to EF, DC to DE, etc., EA is similarly divided in D, DC in A, etc. It follows also, that the greater segment of a line so divided will be itself similarly divided, if a part be cut off from it equal to the less; and that by adding to the whole line its greater segment, another line will be obtained, which is similarly divided.

PROP. XI.—THEOR.—*In an obtuse-angled triangle, the square of the greatest side exceeds the squares of the other two, by twice the rectangle contained by either of the last-mentioned sides, and its continuation to meet a perpendicular drawn to it from the opposite angle.*

Let ABC be a triangle, having the angle ACB obtuse; and let AD be perpendicular to BC produced; the square of AB is equivalent to the squares of AC and CB, and twice the rectangle BC.CD.

Because the straight line BD is divided into two parts in the point C, the square of BD is equivalent (II. 4) to the squares of BC and CD, and twice the rectangle BC.CD. To each of these equivalents add the square of DA; and the squares of DB, DA are equivalent to the squares of BC, CD, DA, and twice the rectangle BC.CD. But, because the angle D is a right angle, the square of BA is equivalent (I. 24, cor. 1) to the squares of BD, DA, and the square of CA is equivalent to the squares of

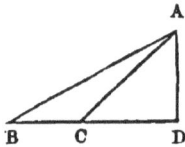

CD, DA; therefore the square of BA is equivalent to the squares of BC, CA, and twice the rectangle BC.CD. Therefore, in an obtuse-angled triangle, etc.

PROP. XII.—THEOR.—*In any triangle, the square of a side subtending an acute angle is less than the squares of the other sides, by twice the rectangle contained by either of those sides, and the straight line intercepted between the acute angle and the perpendicular drawn to that side from the opposite angle.*

Let ABC (see this figure and that of the foregoing proposition) be any triangle, having the angle B acute; and let AD be perpendicular to BC, one of the sides containing that angle; the square of AC is less than the squares of AB, BC, by twice the rectangle CB.BD.

The squares of CB, BD are equivalent (II. 7) to twice the rectangle contained by CB, BD, and the square of DC. To each of these equals add the square of AD; therefore the squares of CB, BD, DA are equivalent to twice the rectangle CB.BD, and the squares of AD, DC. But, because AD is perpendicular to BC, the square of AB is equivalent (I. 24, cor. 1) to the squares of BD, DA, and the square of AC to the squares of AD, DC; therefore the squares of CB, BA are equivalent to the square of AC, and twice the rectangle CB.BD; that is, the square of AC alone is less than the squares of CB, BA, by twice the rectangle CB.BD.

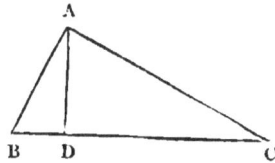

If the side AC be perpendicular to BC, then BC is the straight line between the perpendicular and the acute angle at B; and it is manifest that the squares of AB, BC are equivalent (I. 24, cor. 1) to the square of AC and twice the square of BC. Therefore, in any triangle, etc.

Scho. By means of this or the foregoing proposition, the area of a triangle may be computed, if the sides be given in numbers. Thus, let $AB = 17$, $BC = 28$, and $AC = 25$. From $AB^2 + BC^2$ take AC^2; that is, from $17^2 + 28^2$ take 25^2; the remainder 448 is twice CB.BD. Dividing this by 56, twice BC, the quotient 8 is BD. Hence, from either of the triangles ABD,

ACD, we find the perpendicular AD to be 15; and thence the area is found, by taking half the product of BC and AD, to be 210.

The segments of the base are more easily found by means of the 4th corollary to the fifth proposition of this book, in connection with the principle, that *if half the difference of two magnitudes be added to half their sum, the result is the greater; and if half the difference be taken from half the sum, the remainder is the less.* Thus, if 42, the sum of AB and AC, be multiplied by 8, their difference, and if the product, 336, be divided by 28, the sum of the segments of the base, the quotient 12 is their difference. The half of this being added to the half of 28, the sum 20 is the greater segment CD; and being subtracted from it, the remainder 8 is BD.

To prove the principle mentioned above, let AB be the greater, and BC the less of two magnitudes. Bisect AC in D, and make AE equal to BC. Then AD or DC is half the sum, and ED or DB half the difference of AB and BC; and AB the greater is equivalent to the sum of AD and DB, and BC the less is equivalent to the difference of DC and DB.

Cor. Hence, when AP (I. 24, cor. 6) is not perpendicular to BC—the truth of the corollary can be shown by this and preceding propositions.

Hence, if the sides of a triangle be given in numbers, the line AD can be computed. Thus, if AC=11, AB=14, and BC=7, we have AC²+BC²=121+49=170, and 2AD²=98. Then 170 —98=72, the half of which is 36; and 6, the square root of this, is CD.

PROP. XIII.—PROB.—*To describe a square that shall be equivalent to a given rectilineal figure.*

Let A be the given rectilineal figure; it is required to describe a square that shall be equal to it.

Describe (II. 5 scho.) the rectangle BD equal to A. Then, if the sides of it, BE, ED, be equal to one another, it is a square, and what was required is done. But if they be not equal, produce one of them BE to F, and make EF equal to ED; bisect BF in

G, and from the center G, at the distance GB, or GF, describe
(I. post. 3) the semicircle BHF;
produce DE to H, and join GH.
Therefore, because the straight
line BF is divided equally in G,
and unequally in E, the rectangle
BE.EF, and the square of EG,
are equivalent (II. 5) to the
square of GF, or of GH, because GH is equal to GF. But the
squares of HE, EG are equal (I. 24, cor. 1) to the square of GH;
therefore the rectangle BE.EF and the square of EG are equivalent to the squares of HE, EG. Take away the square of EG,
which is common, and the remaining rectangle BE.EF is equivalent to the square of EH. But the rectangle contained by
BE, EF is the parallelogram BD, because EF is equal to ED;
therefore BD is equivalent to the square of EH; but BD is
equivalent to the figure A; therefore the square of EH is
equivalent to A. The square described on EH, therefore, is the
required square.

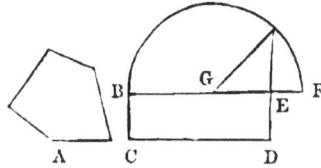

Scho. This is a particular case of the sixteenth proposition
of the fifth book, and an easy solution can be effected by means
of Proportion.

Prop. XIV.—Theor.—*The sum of the squares of the sides
of a trapezium is equivalent to the sum of the squares of the
diagonals, together with four times the square of the straight
line joining the points of bisection of the diagonals.*

Let ABCD be a trapezium, having its diagonals AC, BD
bisected in E and F, and let EF be joined; the squares of AB,
BC, CD, DA are together equivalent to the
squares of AC, BD, together with four
times the square of EF.

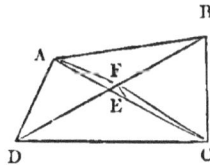

Join AF, FC. The squares of AB, AD
are together equivalent (II. 12, cor.) to
twice the sum of the squares of DF and
AF; and the squares of BC, CD are equivalent to twice the sum of the squares of DF and CF. Add
these equivalents together, and the sum of the squares of AB,
BC, CD, DA is equivalent to four times the square of DF, to-

gether with twice the sum of the squares of AF and CF. But twice the squares of AF and CF are equivalent (II. 12, cor.) to four times the squares of AE and EF; and (II. 4, cor. 2) four times the square of DF is equivalent to the square of BD, and four times the square of AE to the square of AC. Hence the squares of AB, BC, CD, DA are equivalent to the squares of AC and BD, together with four times the square of EF. Therefore, the sum, etc.

Cor. Hence the squares on the diagonals of a parallelogram are together equivalent to the sum of the squares on its sides—for in the case of a parallelogram the line EF vanishes, as the diagonals of a parallelogram bisect each other.

Scho. Hence, if we have the sides and one of the diagonals of a parallelogram in numbers, we can compute the remaining diagonal. Thus, if AB, DC be each $=9$, AD, BC each $=7$, and AC $=8$, we have $AB^2 + BC^2 + CD^2 + DA^2 = 81 + 49 + 81 + 49 = 260$, and $AC^2 = 64$. Taking the latter from the former, and extracting the square root, we find $BD = 14$.

END OF BOOK SECOND.

BOOK THIRD.

ON THE CIRCLE, AND LINES AND ANGLES DE-
PENDING ON IT, AND RECTILINEAL FIGURES
DESCRIBED ABOUT THE CIRCLE.

1. A straight line is said to *touch* a circle, or to be a *tangent*
to it, when it meets the circle, and being produced does not
cut it.

2. Circles are said to *touch* one another, which meet, but do
not cut one another.

3. In a circle, chords are said to be *equally distant from the
center*, when the perpendiculars drawn to them from the center
are equal.

4. And the chord on which the greater perpendicular falls,
is said to be *further from the center.*

5. An *angle in a segment of a circle* is an angle contained
by two straight lines drawn from any point in the arc of the
segment to the extremities of its chord;

6. And an angle is said to *stand upon* the arc intercepted
between the straight lines that contain the angle.

7. A *sector* of a circle is a figure contained by any arc of the
circle, and two radii drawn through its extremities.

8. A *quadrant* is a sector whose radii are perpendicular to
each other. It is easy to show by superposition, that a quadrant is half of a semicircle, and therefore a fourth part of the
entire circle.

9. *Similar segments* of circles are those which contain equal
angles.

10. *Concentric* circles are those which have the same center.

11. A regular polygon is equilateral.

12. When the sides of one rectilineal figure pass through the

5

angular points of another, the figures not coinciding with one another, the interior figure is said to be *inscribed* in the exterior, and the exterior to be *circumscribed*, or *described*, about the interior one.

13. When all the angular points of a rectilineal figure are upon the circumference of a circle, the rectilineal figure is said to be *inscribed* in the circle, and the circle to be *circumscribed*, or *described*, about the rectilineal figure.

14. When each side of a rectilineal figure touches a circle, the rectilineal figure is said to be *circumscribed*, or *described*, about the circle, and the circle to be *inscribed* in the rectilineal figure.

PROPOSITIONS.

PROP. I.—PROB.—*To find the center of a given circle.*

Let ABC be the given circle, and draw any chord AB.

Bisect AB (I. 6) by the perpendicular EDC drawn to the circumference on both sides AB.

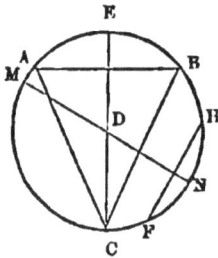

Since EDC bisects AB (const.) and is perpendicular to AB (I. 6), the angles AEB and ACB are also bisected by EDC, and the subtended arcs ACB and AEB (I. 5) are likewise bisected by EDC; therefore the arcs $CA+AE=$ the arcs $CB+BE$; hence $CA+AE$ or CE is the semicircumference, and the perpendicular EDC is a diameter—consequently it passes through the center of the circle.

Then taking any other chord FII, and in same manner it can be shown that the perpendicular MN which bisects the chord is also a diameter, and since all diameters pass through the center of the circle, the point at which they intersect each other, being the only point they have in common, that point is the center of the circle.

Cor. 1. Hence, to find center of any regular polygon (I. 6), bisect the sides by perpendiculars drawn from the points of bisection (I. 7), and the point where the perpendiculars intersect each other is the center of the polygon.

Cor. 2. In a triangle, straight lines drawn from the points of bisection of the three sides to the opposite angles all pass through the same point.

Scho. From the preceding corollary it can be shown that each of the straight lines is divided into two segments at the common point of bisection, of which the segment nearest the angle is double the other.

PROP. II.—THEOR.—*If a straight line drawn from the center of a circle bisect a chord which does not pass through the center, it cuts it at right angles; and* (2) *if it cut it at right angles, it bisects it.*

Let ABC be a circle; and let ED, a straight line drawn from the center E, bisect any chord AB, which does not pass through the center, in the point D; ED cuts AB at right angles.

Join EA, EB. Then in the triangles ADE BDE, AD is equal to DB, DE common, and (I. def. 16) the base EA is equal to the base EB; therefore (I. 4) the angles ADE, BDE are equal; and consequently (I. def. 10) each of them is a right angle; wherefore ED cuts AB at right angles.

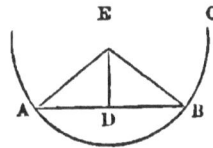

Next, let ED cut AB at right angles; ED also bisects it.

The same construction being made, because the radii EA, EB are (I. def. 16) equal, the angle EAD is equal (I. 1, cor.) to EBD; and the right angles ADE, BDE are equal; therefore in the two triangles EAD, EBD there are two angles in one equal to two angles in the other, each to each, and the side ED, which is opposite to one of the equal angles in each, is common; therefore (I. 14) AD is equal to DB. If a straight line, therefore, etc.

Cor. 1. Hence, in an isosceles triangle, a straight line drawn from the vertex bisecting the base is perpendicular to it; and a straight line drawn from the vertex perpendicular to the base, bisects it.

Cor. 2. Let the straight line AB cut the concentric circles ABC, DEF in the points A, D, E, B; AD is equal to EB, and AE to DB. From the common center G, draw GH perpen-

dicular to AB. Then (III. 2) AH is equal to HB, and DH to
HE. From AH take DH, and from HB
take HE, and the remainders, AD, EB,
are equal. To these equals add DE, and
the sums AE, DB are equal.

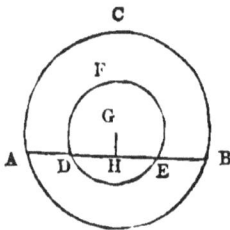

Cor. 3. Any number of parallel chords
in a circle are all bisected by a diameter
perpendicular to them.

PROP. III.—THEOR.—*Two chords of a circle which are not
both diameters, can not bisect each other.*

From the definition of the circle, the center is the only point
in the circle which is equally distant
from all parts of the circumference; then
any chord which passes through the
center is bisected at the center ; the
diameter is the only chord (III. 1)
which passes through the center, therefore any other chord AB can not bisect
a diameter CD. And when neither chord
AB nor EF is a diameter, their point of
intersection not being the center of the
circle, is unequally distant from the circumference ; therefore
the chords AB and EF are not bisected by each other.

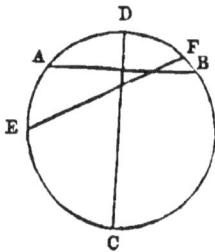

PROP. IV.—THEOR.—*If two circles cut one another, they have
not the same center.*

Let ABC and DBE be two circles which cut one another in
B; they will not have the
same center. For AC is the
diameter of ABC (III. 1), and
DE is the diameter of DBE.
But the intersection of the
two diameters AC and FH
is the center of the circle
ABC (III. 1), and the intersection of the two diameters
LM and DE is the center of the circle DBE. Now the points

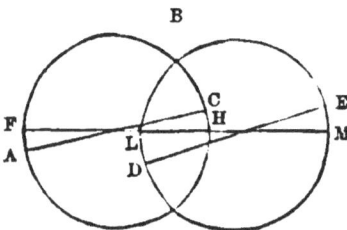

of intersection of these diameters are different, therefore ABC has not the same center with DBE. Wherefore, if two circles, etc.

Cor. 1. Hence, if one circle touches another internally, they have not the same center.

Cor. 2. One circle can not cut another in more than two points, nor touch another in more than one point.

PROP. V.—THEOR.—*If from any point within a circle, which is not the center, straight lines be drawn to the circumference;* (1) *the greatest is that which passes through the center, and* (2) *the continuation of that line to the circumference, in the opposite direction, is the least;* (3) *of others, one nearer to the line passing through the center is greater than one more remote; and* (4) *from the same point there can be drawn only two equal straight lines, one upon each side of either the longest or shortest line, and making equal angles with that line.*

Let ABCD be a circle, E its center, and AD a diameter, in which let any point F be taken, which is not the center; of all the straight lines FA, FB, FC, etc., that can be drawn from F to the circumference, FA is the greatest, and FD the least; and of the others, FB is greater than FC.

1. Join BE, CE. Then (I. 21, cor.) BE, EF are greater than BF; but AE is equal to EB; therefore AF, that is, AE, EF, is greater than BF.

2. Because CF, FE are greater (I. 21, cor.) than EC, and EC is equal to ED; CF, FE are greater than ED. Take away the common part FE, and (I. ax. 5) the remainder CF is greater than the remainder FD.

3. Again: because BE is equal to CE, and FE common to the triangles BEF, CEF; but the angle BEF is greater than CEF; therefore (I. 22) the base BF is greater than the base CF.

4. Make (I. 13) the angle FEH equal to FEC, and join FH. Then, because CE is equal to HE, EF common to the two triangles CEF, HEF, and the angle CEF equal to the angle HEF; therefore (I. 3) the base FC is equal

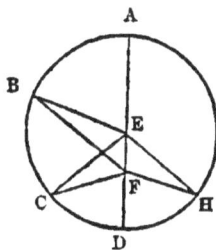

to the base FH, and the angle EFC to the angle EFH. But, besides FH, no other straight line can be drawn from F to the circumference equal to FC (I. ax. 9).

PROP. VI.—THEOR.—*If from any point without a circle straight lines be drawn to the circumference;* (1) *of those which fall upon the concave part of the circumference, the greatest is that which passes through the center ; and* (2) *of the rest, one nearer to the greatest is greater than one more remote.* (3) *But of those which fall upon the convex part, the least is that which when produced passes through the center; and* (4) *of the rest, one nearer to the least is less than one more remote. And* (5) *only two equal straight lines can be drawn from the point to either part of the circumference, one upon each side of the line passing through the center, and making equal angles with it.*

Let ABF be a circle, M its center, and D any point without it, from which let the straight lines DA, DE, DF be drawn to the circumference. Of those which fall upon the concave part of the circumference AEF, the greatest is DMA, which passes through the center ; and a line DE nearer to it is greater than DF, one more remote. But of those which fall upon the convex circumference LKG, the least is DG, the external part of DMA ; and a line DK nearer to it is less than DL, one more remote.

1. Join ME, MF, ML, MK ; and because MA is equal to ME, add MD to each; therefore AD is equal to EM, MD ; but (I. 21, cor.) EM, MD are together greater than ED ; therefore, also, AD is greater than ED.

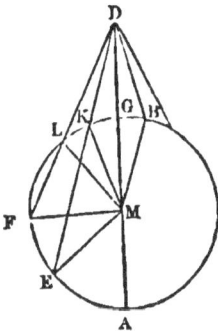

2. Because ME is equal to MF, and MD common to the triangles EMD, FMD, but the angle EMD is greater than FMD ; therefore (I. 22) the base ED is greater than the base FD.

3. Because (I. 21, cor.) MK, KD are greater than MD, and MK is equal to MG, the remainder KG is greater (I. ax. 5) than the remainder GD ; that is, GD is less than KD.

4. Because MK is equal to ML, and MD common to the triangles KMD, LMD, but the angle DMK less than DML; therefore the base DK is less (I. 22) than the base DL.

5. Make (I. 13) the angle DMB equal to DMK, and join DB. Then, because MK is equal to MB, MD common to the triangles KMD, BMD, and the angle KMD equal to BMD; therefore (I. 3) the base DK is equal to the base DB, and the angle MDK to the angle MDB. But, besides DB, there can be no straight line drawn from D to the circumference equal to DK (I. ax. 9).

Prop. VII.—Theor.—*Equal chords in a circle are equally distant from the center ; and* (2) *chords which are equally distant from the center are equal to one another.*

Let the chords AB, CD, in the circle ABDC, be equal to one another; they are equally distant from the center.

Take (III. 1) E the center of the circle, and draw (I. 7) EF, EG perpendiculars to AB, CD; join also EA, EC. Then, because the straight line EF, passing through the center, cuts the chord AB, which does not pass through the center, at right angles, it also (III. 2) bisects it ; wherefore AF is equal to FB, and AB is double of AF. For the same reason, CD is double of CG; but AB is equal to CD ; therefore AF is equal (I. ax. 7) to CG. Then, in the right-angled triangles EFA, EGC, the sides EA, AF are equal to the sides EC, CG, each to each, therefore (I. 3) the sides EF, EG are equal. But chords in a circle are said (III. def. 3) to be equally distant from the center, when the perpendiculars drawn to them from the center are equal; therefore AB, CD are equally distant from the center.

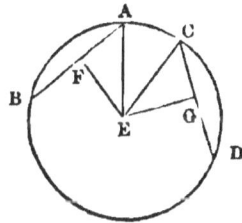

Next, if the chords AB, CD be equally distant from the center, that is, if FE be equal to EG, AB is equal to CD. For, the same construction being made, it may, as before, be demonstrated that AB is double of AF, and CD of CG ; and because the right-angled triangles EFA, EGC have the sides AE, EF equal to CE, EG, each to each, the sides AF, CG are also (I. 3)

equal to one another. But AB is double of AF, and CD of CG; wherefore AB is equal (I. ax. 6) to CD. Therefore equal chords, etc.

Cor. Hence, the diameter of a circle is the greatest chord; (2) of others, one nearer to the center is greater than one more remote; and (3) the greater is nearer to the center than the less.

PROP. VIII.—THEOR.—*The straight line drawn perpendicular to a diameter of a circle, through its extremity, falls without the circle ; but any other straight line drawn through that point cuts the circle.*

Let ABC be a circle, of which D is the center, and AB a diameter; if AE be drawn through A perpendicular to AB, it falls without the circle.

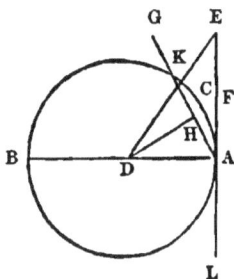

In AE take any point F; and draw DF, meeting the circumference in C. Because DAF is a right angle, it is greater (I. 20) than DFA; and therefore (I. 21) DF is greater than DA. But (I. def. 16) DA is equal to DC; therefore DF is greater than DC, and the point F is therefore without the circle; and in the same manner it may be shown, that any other point in AF, except the point A, is without the circle.

Again: any other straight line drawn through A cuts the circle.

Let AG be drawn in the angle DAF, and draw (I. 8) DH perpendicular to AG, and meeting the circumference in C. Then, because DHA is a right angle, and DAH less than a right angle, being a part of DAE, the side DH is less (I. 20) than the side DA. But (I. def. 16) DK is equal to DA; therefore DH is less than DK; the point H is therefore within the circle; and AG cuts the circle, since its continuation through A must fall on the opposite side of EAL, and must therefore be without the circle. Therefore the straight line, etc.

Cor. From this it is manifest that the straight line which is drawn at right angles to a diameter of a circle from its extremity touches (III. def. 1) the circle; and that it touches it only in one point, because at every point except A, it falls without the circle. It is also evident, that there can be but one tangent at the same point of a circle.

PROP. IX.—PROB.—*From a given point, either without a given circle, or in its circumference, to draw a straight line touching the circle.*

First: let A be a given point without the given circle BCD; it is required to draw from A a straight line touching the circle.

Find (III. 1) E the center of the circle, and draw AE cutting the circumference in D; from the center E, at the distance EA, describe (I. post. 3) the circle AFG; from D draw (I. 7) DF at right angles to EA; and draw EBF, and join AB. AB touches the circle BCD.

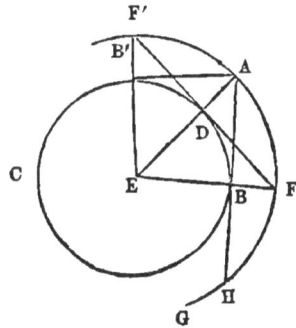

Because E is the center of the circles, EA is equal to EF, and ED to EB; therefore the two sides AE, EB are equal to the two FE, ED, each to each, and they contain the angle AEF common to the two triangles AEB, FED; therefore the angle EBA is equal (I. 3) to EDF, and is, therefore, a right angle, because (const.) EDF is a right angle. Now, since EB is drawn from the center, it is part of a diameter of which B is one extremity; but a straight line drawn from the extremity of a diameter at right angles to it touches (III. 8, cor.) the circle; therefore AB touches the circle; and it is drawn from the given point A; which was to be done.

Secondly: if the given point be in the circumference of the circle, as the point D, draw DE to the center E, and DF at right angles to DE; DF touches (III. 8, cor.) the circle.

Cor. 1. If AB be produced to H, AH is bisected (III. 2) in B. Hence a chord in a circle touching a concentric one is bisected at the point of contact.

Scho. 1. It is evident that from any point A without the circle, two tangents may be drawn to the circle, and that these are equal to one another, being equal respectively to the equal lines DF and DF′.

Scho. 2. The construction of the first case of this problem is as easily effected in practice, by describing a circle on AE as diameter, as its circumference will cut that of the given circle in the points B and B′. The reason of this will be evident from the twelfth proposition of this book.

Cor. 2. Hence in a given straight line AB, a point may be found such that the difference of its distances from two given points C, D, may be equal to a given straight line. Join CD, and from D as center, with a radius, DE equal to the given difference, describe a circle; draw CFG perpendicular to AB, and make FG equal to FC; through C, G describe a circle touching the other circle; join B, D, the centers of the two circles, and draw BC; BC, BD are evidently the required lines.

Cor. 3. In the same manner, if a circle were described from D as center, with the sum of two given lines as radius, and a circle were described through C and G touching that circle, straight lines drawn from the center of that circle to C and D would be equal to the given sum.

PROP. X.—THEOR.—*The angle at the center of a circle is double of the angle at the circumference, upon the same base, that is, upon the same part of the circumference.*

In the circle ABC, let BEC be an angle at the center, and BAC an angle at the circumference, which have the same arc BC for their base; BEC is double of BAC.

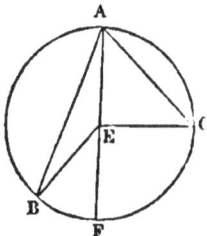

Draw AE, and produce it to F; and first, let E, the center of the circle, be within the angle BAC. Because EA is equal to EB, the angle EAB is equal (I. 1, cor.) to EBA; therefore the angles EAB, EBA are together double of EAB; but (I. 20) the angle BEF is equal to EAB, EBA; therefore also BEF is double of EAB.

For the same reason, the angle FEC is double of the angle
EAC; therefore the whole angle BEC is double of the whole
BAC.

Again: let E the center of the circle be without the angle
BAC; it may be demonstrated, as in the
first case, that the angle FEC is double of
FAC, and that FEB, a part of the first, is
double of FAB a part of the other; therefore the remaining angle BEC is double of
the remaining angle BAC. The angle at
the center, therefore, etc.

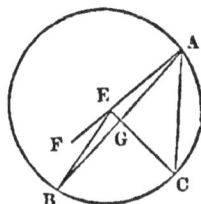

Scho. That *if two magnitudes be double
of two others, each of each, the sum and difference of the first
two are respectively double of the sum and difference of the
other two.* It is thus proved by Playfair: "Let A and B, C
and D be four magnitudes, such that $A=2C$, and $B=2D$;
then $A+B=2(C+D)$. For since $A=C+C$, and $B=D+D$,
adding equals to equals, $A+B=(C+D)+C+D=2(C+D)$.
So, also, if A be greater than B, and therefore C greater than
D, since $A=C+C$, and $B=D+D$, taking equals from equals,
$A-B=(C-D)+(C-D)$, that is $A-B=2(C-D)$."

The following is an outline of another proof of the second
case: from the triangle BGE we have (I. 20) $BGC=BEC+B$
$=BEC+EAG$ (I. 1). We have also from the triangle ACG,
in a similar manner, $BGC=BAC+C=BAC+EAC=2BAC+$
$EAG.$ Hence (I. ax. 1) $BEC+EAG=2BAC+EAG.$ Take
away EAG, etc. If in the first diagram, CE were produced to
meet AB, the first case might be proved in a similar manner.
The second case might also be proved by drawing from C to
AB a straight line meeting it in a point H, and making with
AC an angle equal to BAC. Then, by taking the difference
between the equal angles EAC, ECA, and the equal ones HAC,
HCA, we have EAB and ECH equal, and therefore EBA=
ECH. But EGB=HGC; and therefore (I. 20, cor. 5) BEC=
BHC=2BAC. The same proof, with some obvious variations,
would be applicable in the first case.

There is evidently a third case, viz., when AB or AC passes
through the center; but though this case is not given in a separate form, its proof is contained in that of either of the others.

PROP. XI.—THEOR.—*In a circle*, (1) *the angle in a semicircle is a right angle;* (2) *the angle in a segment greater than a semicircle is acute;* and (3) *the angle in a segment less than a semicircle is obtuse.*

Let ABC be a circle, of which F is the center, BC a diameter, and consequently BAC a semicircle; and let the segment BAD be greater, and BAE less than a semicircle; the angle BAC in the semicircle is a right angle; but the angle BAD in the segment greater than a semicircle is acute; and the angle BAE in the segment less than a semicircle is obtuse.

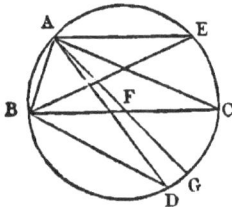

Draw AF and produce it to G. Then (III. 10) the angle BAG at the circumference, is half of BFG at the center, both standing on the same arc BG; and for the same reason, GAC is half of GFC. Therefore the whole angle BAC is half of the angles BFG, GFC; and (I. 9) these are together equivalent to two right angles; therefore DAC is a right angle, and it is an angle in a semicircle. But (I. ax. 9) the angle BAD is less, and BAE greater than the right angle BAC; therefore an angle in a segment greater than a semicircle is acute, and an angle in a segment less than a semicircle is obtuse.

PROP. XII.—THEOR.—*If a straight line touch a circle, the straight line drawn from the center to the point of contact is perpendicular to the line touching the circle.*

Let the straight line FH touch the circle ABCD at the point C; the straight line CA drawn from that point to the center of the circle is perpendicular to the line touching the circle. Draw a diameter BD parallel to FH (I. 18). and with the diameter BD as a base, and the point C as a vertex, make the triangle BCD. But BCD is a right angle (III. 11). FCB and HCD are equivalent to a right angle (I. 9), and because BD is parallel to FH (const.), FCB is equal to CBD, and HCD is equal to CDB (I. 15, cor. 2); and

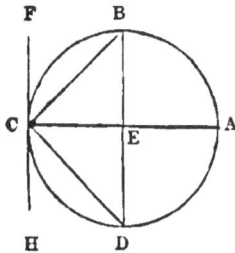

at the point C draw a perpendicular CA to FH (I. 7), it will also be perpendicular to BD (I. 17), then FCB and BCE are equivalent to BCD (I. ax. 1); taking away the common angle BCE, we have FCB equal to ECD; but FCB is equal to CBD, hence ECD is equal to CBD; and in same manner it can be shown that ECB is equal to CDB. The triangles EBC and EDC having EC common, the angle EBC equal to the angle ECD, and the angles BEC and CED both right angles, are equal (I. 3), therefore EB is equal to ED, and the diameter BD is bisected by CA, then CA passes through the center of the circle (III. 1). Wherefore, if a straight line, etc.

Cor. Hence, *conversely*, if a straight line touch a circle, a straight line drawn from the point of contact, perpendicular to the tangent, passes through the center.

Prop. XIII.—Theor.—*If one circle touch another internally or externally in any point, the straight line which joins their centers being produced, passes through that point.*

Let *ab*C and DEC be two circles which touch one another internally, their centers will be in the same straight line with their point of contact.

At the point of contact C draw the tangent FH, and at this point erect a perpen-dicular C*a*E (I. 7). The diameter of *ab*C is perpendicular to the tangent at the point of contact (III. 12), hence C*a* is the diameter of *ab*C, and passes through the center of *ab*C (III. 1). Now FH

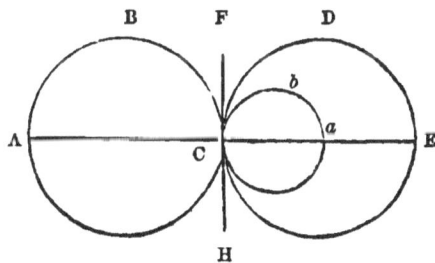

is also tangent to the circle DEC (const.) at the point C; hence, the perpendicular CE (III. 12) is the diameter of DEC, and passes (III. 1) through the center of DEC. But C*a* and CE are in the same straight line, hence the centers of the circles *ab*C and DEC, and the point of contact C, are in the straight line C*a*E. Wherefore, if one circle touch, etc.

Or, let ABC and CDE be two circles which touch one another externally in the point C; their centers will be in the same straight line with the point of contact.

At the point C draw a tangent FH, which will be perpendicular to the diameter of ABC (III. 12). FH being tangent at the point of contact of the circles, is also perpendicular to the diameter of CDE (III. 12). ACF is a right angle (I. def. 10), and FCE for the same reason is a right angle, and both equal to one another (I. def. 10, and ax. 11), hence are two right angles; then (I. 10) AC and CE form the same straight line; but AC passes through the center of ABC, and CE passes through the center of CDE, being their respective diameters (III. 1), therefore the centers of the circles are in the same straight line with the point C. Wherefore, if two circles touch each other, etc.

PROP. XIV.—THEOR.—*Similar segments of circles upon equal bases are equal to one another, and have equal arcs.*

Let AEB, CFD be similar segments of circles upon the equal straight lines or bases, AB, CD; the segments are equal; and likewise the arcs AEB, CFD are equal.

For if the segment AEB be applied to the segment CFD, so

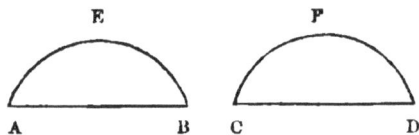

that the point A may be on C, and the straight line AB on CD, the point B will coincide with D, because AB is equal to CD; therefore the straight line AB coinciding with CD, the segment AEB must coincide (I. def. 16) with the segment CFD, and is therefore equal (I. ax. 8) to it; and the arcs AEB, CFD are equal, because they coincide. Therefore similar segments, etc.

PROP. XV.—PROB.—*A segment of a circle being given, to complete the circle of which it is a segment.*

Assume three points in the arc of the segment, and find (III. 1) the center of the circle. From that center, at the distance between it and any point in the arc describe a circle, and it will evidently be the one required.

PROP. XVI.—THEOR.—*In equal circles, or in the same circle, equal angles stand upon equal arcs, whether they are at the centers or the circumferences.*

Let ABC, DEF be equal circles, having the equal angles BGC, EHF at their centers, and BAC, EDF at their circumferences; the arc BKC is equal to the arc ELF.

Join BC, EF; and because the circles ABC, DEF are equal, their radii are equal; therefore the two sides BG, GC are equal

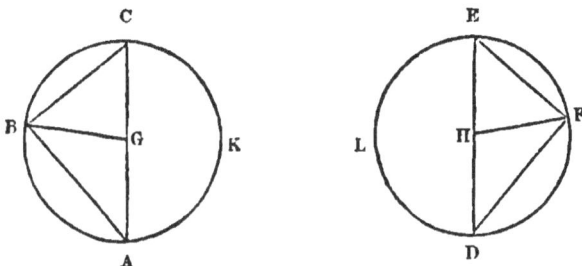

to the two, EH, HF; and (hyp.) the angles G and H are equal; therefore (I. 3) the base BC is equal to the base EF. Then, because the angles A and D are equal, the segment BAC is similar (III. def. 8) to the segment EDF; and they are upon equal straight lines BC, EF; but (III. 14) similar segments of circles upon equal straight lines have equal arcs; therefore the arc BAC is equal to the arc EDF. But the whole circumference ABC is equal to the whole DEF, because the circles are equal; therefore the remaining arc BKC is equal (I. ax. 3) to the remaining arc ELF. Wherefore, in equal circles, etc.

Cor. 1. Conversely, in equal circles, or in the same circle, the angles which stand upon equal arcs are equal to one another, whether they are at the centers or the circumferences. (I. def. 19.)

Cor. 2. Hence, in a circle, the arcs intercepted between parallel chords are equal. For if a straight line be drawn transversely, joining two extremities of the chords, it will (I. 16) make equal angles with the chords; and therefore the arcs on which these stand are equal.

Cor. 3. Hence, in equal circles, equal chords divide the circumferences into parts which are equal, each to each.

Prop. XVII.—Prob.—*To bisect a given arc of a circle.*

Let ADB be a given arc; it is required to bisect it.

Draw AB, and (I. 6 and 7) bisect it in C, by the perpendicular CD; the arc ABD is bisected in the point D.

Join AD, DB. Then, because AC is equal to CB, CD common to the triangles ACD, BCD, and the angle ACD equal to BCD, each of them being a right angle; therefore (I. 3) AD is equal to BD. But (III. 16, cor. 3), in the same circle, equal lines cut off equal arcs, the greater equal to the greater, and the less to the less; and AD, DB are each of them less than a semicircle, because (III. 1) DC, or DC produced, passes through the center; wherefore the arc AD is equal to the arc DB; therefore the given arc is bisected in D; which was to be done.

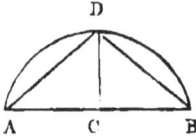

Prop. XVIII.—Theor.—*If a straight line touch a circle, and from the point of contact a straight line be drawn dividing the circle into two segments ; the angles made by this line with the tangent are equivalent to the angles which are in the alternate segments.*

Let the straight line DE touch the circle BAG in the point B, and let the straight line BA be drawn dividing the circle into the segments AGB, AB ; the angle ABE is equal to any angle in the segment AGB, and the angle ABD to any angle in AB.

If AB (fig. 1) be perpendicular to DE, it passes (III. 12, cor.) through the center, and the segments being therefore semicircles, the angles in them are (III. 11) right angles, and consequently equal to those which AB makes with DE.

But if BA (fig. 2) be not perpendicular to DE, draw BF perpendicular to it ; join FA and produce it to E; join also CA,

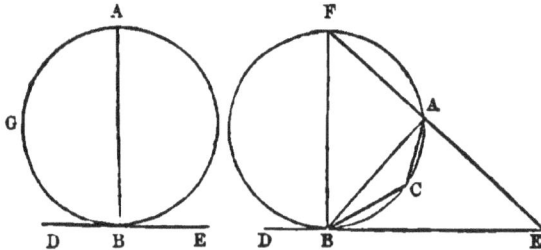

CB, C being any point in the arc ACB. Then (III. 12, cor.)
BF is a diameter, and (III. 11, and I. 9) the angles BAF, BAE
are right angles. Therefore, in the triangles BAE, FBE, the
angle E is common, and the angles BAE, FBE equal, being

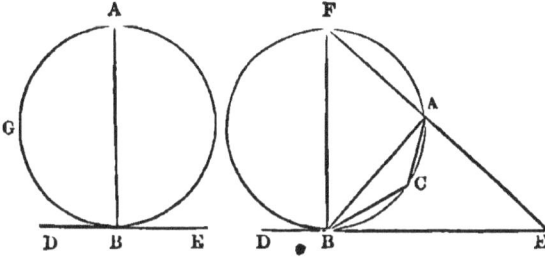

right angles; wherefore (I. 20, cor. 5) the remaining angle
ABE is equal to the remaining angle F, which is an angle in
the remote or alternate segment BFA.

Again: the two angles ABD, ABE are equal (I. 9) to two
right angles; and because ACBF is a quadrilateral in the circle,
the opposite angles C and F are also equivalent (I. 20, cor. 1)
to two right angles; therefore (I. ax. 1) the angles ABD, ABE
are together equal to C and F. From these equals take away
the angles ABE and F, which have been proved to be equal;
then (I. ax. 3) the remaining angle ABD is equal to the remaining angle C, which is an angle in the remote segment
ACB. If, therefore, a straight line, etc..

Scho. 1. The first case is wanting in most editions of Euclid.
In the second diagram, FA and DB will meet (I. 19) if produced, the angles FBE, BFA together being evidently less
than two right angles.

Scho. 2. Let AB be a fixed chord, and
through B draw any other line DE cutting the circle in C; and join AC. Now
(III. 16, cor. 1), wherever C is taken in
the arc ACB, the angle ACE is con-
stantly of the same magnitude; and so
also (I. 9) is the exterior angle ACD.
If C be now taken as coinciding with B,
the straight line DE will become the
tangent D'BE', AC will coincide with AB, and the angles

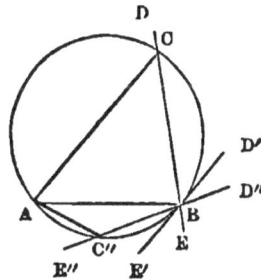

ACE, ACD will become ABE′, ABD′. If again C take the position C″, the angles ACE, ACD will become AC″E″, AC″D″. Now, the equality of ABE′, ACE, and of ABD′, AC″D″ is what is proved in the eighteenth proposition, and from the equality of ACD and AC″D″ cor. 2 follows by the addition of ACB.

Cor. 1. Angles in the same segment of a circle are equal to one another.

Cor. 2. The opposite angles of any quadrilateral figure described in a circle are together equivalent to two right angles; and *conversely*, if two opposite angles of a quadrilateral be together equal to two right angles, a circle may be described about it.

Cor. 3. If the circumference of a circle be cut by two straight lines which are perpendicular to one another, the squares of the four segments between the point of intersection of the two lines and the points in which they meet the circumference, are together equal to the square of the diameter.

PROP. XIX.—PROB.—*Upon a given straight line, to describe a segment of a circle containing an angle equal to a given angle.*

Let AB be the given straight line, and C the given angle; it is required to describe on AB a segment of a circle containing an angle equal to C.

First: if C be a right angle, bisect (I. 6) AB in F, and from the center F, at the distance FB, describe the semicircle AHB; therefore (III. 11) any angle AHB in the semicircle is equal to the right angle C.

But if C be not a right angle, make (I. 13) the angle BAD equal to C, and (I. 7) from A draw AE perpendicular to AD; bisect (I. 5 and 6) AB by the perpendicular FG, and join GB. Then, because AF is equal to FB, FG common to the triangles AFG, BFG, and the angle AFG equal to BFG, therefore (I. 3) AG is equal to GB; and the circle described from the center G, at the distance GA, will pass through the point B; let this be the circle AHB. Then, because from the point A, the extremity

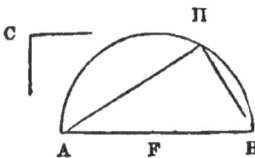

of the diameter AE, AD is drawn at right angles to AE, AD (III. 8, cor.) touches the circle; and (III. 18) because AB, drawn from the point of contact A, cuts the circle, the angle DAB is equal to any angle in the alternate segment AHB; but DAB is equal to C; therefore also C is equal to any angle in the segment AHB; wherefore upon the given straight line AB the segment AHB is described, which contains an angle equal to C; which was to be done.

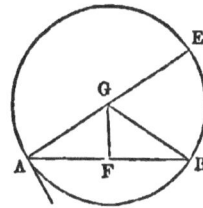

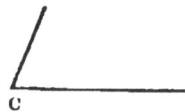

Scho. It is evident there may be two segments answering the conditions of the problem, one on each side of the given line. It is also plain, that when C is an acute angle, and the segment is to be above AB, G is above AB; but when obtuse, it is below it. It is likewise plain, that the angle BAE is the complement of the given angle C; that is, the difference between it and a right angle.

Cor. Hence from a given circle can be cut off a segment which shall contain an angle equal to a given angle.

PROP. XX.—THEOR.—*If two chords of a circle cut one another, the rectangle contained by the segments of one of them is equal to the rectangle contained by the segments of the other.*

In the circle LBM, let the two chords LM, BD cut one another in the point F; the rectangle LF.FM is equal to the rectangle BF.FD.

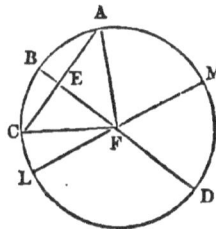

If LM, BD both pass through the center, so that F is the center, it is evident that LF, FM, BF, FD, being (I. def. 16) all equal, the rectangle LF.FM is equal to the rectangle BF.FD.

But let one of them, BD, pass through the center and cut the other, AC, which does not pass through the center, at right angles, in the point E. Then, if BD be bisected in F, F is the center. Join AF; and because BD which passes through the center, is perpendicular to AC, AE, EC are (III. 2) equal to one another. Now, because BD is

divided equally in F, and unequally in E, the rectangle BE.ED,
and the square of EF are equivalent (II. 5) to the square of FB;
that is (I. 23, cor. 2), to the square of FA. But (I. 24, cor. 1)
the squares of AE, EF are equivalent to the square of FA;
therefore the rectangle BE.ED and the square of EF are
equivalent to the squares of AE, EF. Take away the common
square of EF, and the remaining rectangle BE.ED is equivalent
to the remaining square of AE; that is, to the rectangle
AE.EC.

Next, let BD pass through the center, and cut AC, which
does not pass through the center, in E, but not at right angles.
Then, as before, if BD be bisected in F, F is the center of the
circle. Join AF, and (I. 8) draw FG per-
pendicular to AC; therefore (III. 2) AG is
equal to GC; wherefore (II. 5) the rectangle AE.EC and the square of EG are equivalent to the square of AG. To each of
these equivalents add the square of GF;
therefore the rectangle AE.EC and the
squares of EG, GF are equivalent to the
squares of AG, GF; but (I. 24, cor. 1) the squares of EG, GF are
equivalent to the square of EF; and the squares of AG, GF are
equivalent to the square of AF; therefore the rectangle AE.EC
and the square of EF are equivalent to the square of AF; that
is (I. 23, cor. 2), to the square of FB. But (II. 5) the square
of FB is equivalent to the rectangle BE.ED, together with the
square of EF; therefore the rectangle AE.EC and the square
of EF are equivalent to the rectangle BE.ED and the square
of EF; take away the common square of EF, and the remaining rectangle AE.EC is equivalent to the remaining rectangle
BE.ED.

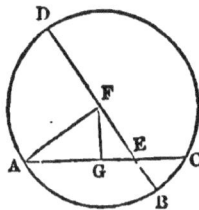

Lastly: let neither of the lines pass through the center, and
through E, the point of intersection, draw a diameter. Then,
the rectangle AE.EC is equivalent, as has been shown, to the
rectangle DE.EB; and, for the same reason, the rectangle of
the other chord is equivalent to the same rectangle DE.EB;
therefore (I. ax. 1) the rectangle AE.EC is equivalent to the
rectangle of the other chord. If, therefore, two chords of a
circle, etc.

Scho. The second and third cases may be thus demonstrated in one:

Join FC. Then the rectangle AE.EC is equivalent (II. 5, cor. 5) to the difference of the squares of AF and FE, or of DF and FE, or (II. 5, cor. 1) to the rectangle DE.EB.

Proportion, however, affords much the easiest method of demonstrating both this proposition and the following.

PROP. XXI.—THEOR.—*If from any point without a circle two straight lines be drawn, one of which cuts the circle, and the other touches it ; the rectangle contained by the whole line which cuts the circle, and the part of it without the circle, is equivalent to the square of the line which touches it.*

Let D be any point without the circle ABC, and DCA, DB two straight lines drawn from it, of which DCA cuts the circle in C and A, and DB touches it in B; the rectangle AD.DC is equivalent to the square of DB.

There are two cases—first : let DCA pass through the center E, and join EB. Therefore (III. 12) the angle EBD is a right angle; and (II. 6) because the straight line AC is bisected in E, and produced to D, the rectangle AD.DC and the square of EC are together equivalent to the square of ED ; and CE is equal to EB; therefore the rectangle AD.DC and the square of EB are equivalent to the square of ED. But (I. 24, cor. 1) the square of ED is equivalent to the squares of EB, BD, because EBD is a right angle ; therefore the rectangle AD.DC and the square of EB are equivalent to the squares of EB, BD ; take away the common square of EB, and the remaining rectangle AD.DC is equivalent to the square of the tangent DB.

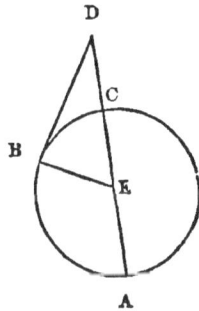

Second : if DCA do not pass through the center, take (III. 1) the center E, and draw (I. 8) EF perpendicular to AC, and join EB, EC, ED. Then, because the straight line EF, which passes through the center, is perpendicular to the chord AC, AF is equal (III. 2) to FC. And (II. 6) because AC is bisected in F, and produced to D, the rectangle AD.DC and the square of FC are equivalent to the square of FD. To each of these equals add the square of FE; therefore the rectangle AD.DC, and the squares of CF, FE are equivalent to the squares of DF, FE; but (I. 24, cor. 1) the square of ED is equivalent to the squares of DF, FE, because EFD is a right angle; and the square of EC, or (I. 23, cor. 2) of EB, is equivalent to the squares of CF, FE; therefore the rectangle AD.DC and the square of EB are equivalent to the square of ED. But (I. 24, cor. 1) the squares of EB, BD are equivalent to the square of ED, because EBD is a right angle; therefore the rectangle AD.DC and the square of EB are equivalent to the squares of EB.BD. Take away the common square of EB; therefore the remaining rectangle AD.DC is equivalent to the square of DB; wherefore, if from any point, etc.

Scho. The second case may be demonstrated more briefly thus: Join AE. Then the rectangle AD.DC is equal (II. 5, cor. 5) to the difference of the squares of ED and EC, or of ED and EB, or (I. 24, cor. 1) to the square of DB.

Cor. 1. If from any point without a circle there be drawn two straight lines cutting it, as AB, AC, the rectangles contained by the whole lines and the parts of them without the circle are equivalent to one another, viz., the rectangle BA.AE to the rectangle CA.AF; for each rectangle is equivalent to the square of the tangent AD.

Cor. 2. If two straight lines intersect each other, so that the rectangle under the segments of one of them is equal to the rectangle under the segments of the other, their extremities lie in the circumference of a circle.

Scho. By means of the first corollary, it would be shown in a similar manner, that if two straight lines meet in a point, and if they be so divided that the rectangle under one of them and its segment next the common point is equal to the rectangle

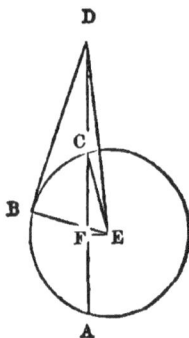

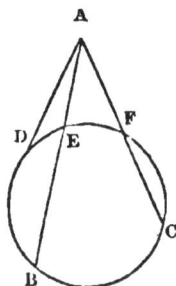

under the other and its corresponding segment, the points of section and the extremities remote from the common point lie in the circumference of the same circle.

PROP. XXII.—THEOR.—*If from a point without a circle there be drawn two straight lines, one of which cuts the circle, and the other meets it ; and if the rectangle contained by the whole line which cuts the circle, and the part of it without the circle, be equivalent to the square of the line which meets it, the line which meets the circle touches it.*

If from a point without the circle ABC two straight lines DCA and DB be drawn, of which DCA cuts the circle, and DB meets it ; and if the rectangle AD.DC be equivalent to the square of DB ; DB touches the circle.

Draw (III. 9) the straight line DE touching the circle ABC ; find (III. 1) the center F ; and join FE, FB, FD. Then (III. 12) FED is a right angle ; and (III. 21) because DE touches the circle ABC, and DCA cuts it, the rectangle AD.DC is equivalent to the square of DE. But (hyp.) the rectangle AD.DC is equivalent to the square of DB ; therefore the square of DE is equivalent to the square of DB ; and the straight line DE equal (I. 23, cor. 3) to the straight line DB. But FE is equal to FB, and the base FD is common to the two triangles DEF, DBF ; therefore (I. 4) the angle DEF is equal to DBF ; but DEF is a right angle ; therefore, also, DBF is a right angle ; and FB is a part of a diameter, and the straight line which is drawn at right angles to a diameter, from its extremity, touches (III. 8, cor.) the circle ; therefore DB touches the circle ABC. Wherefore, if from a point, etc.

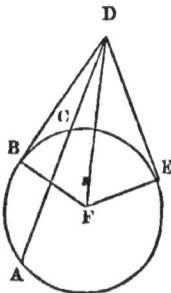

PROP. XXIII.—PROB.—*To divide a given straight line into two parts, such that the square of one of them may be equivalent to the rectangle contained by the other, and a given straight line.*

Let AB, AC be two given straight lines ; it is required to divide AB into two parts, such that the square of one of them may be equivalent to the rectangle under AC and the other.

On CB, the sum of the given lines, describe the semicircle CDB, and draw AD perpendicular to CB; bisect CA in E, and join DE; and make EF equal to ED; then the square of AF is equivalent to the rectangle AC.FB.

Describe the semicircle CGA, cutting ED in G, and join GF.

Then, because FE, EG are respectively equal to DE, EA, and the angle FED common, GF is equal to AD, and the angle EGF to EAD, which is a right angle, and therefore GF touches the circle CGA. Hence (III. 20) the rectangle CA.AB is equivalent to the square of AD, or of FG, or (III. 21) to the rectangle CF.FA. But (II. 1) the rectangle CA.AB is equivalent to the two CA.AF, CA.FB, and (II. 3) the rectangle CF.FA is equivalent to CA.AF and the square of AF. From these equivalents take away the rectangle CA.AF, and there remains the rectangle CA.FB equivalent to the square of AF.   •

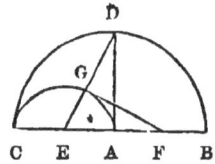

Scho. The tenth proposition of the second book is the particular case of this problem in which the given lines are equal.

PROP. XXIV.—PROB.—*To draw a common tangent to two given circles.*

Let BDC, FHG be given circles; it is required to draw a common tangent to them.

Join their centers A, E, and make BK equal to FE; from A

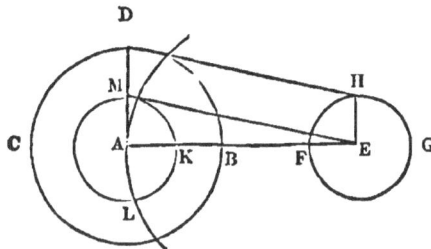

as center, with AK as radius, describe a circle cutting another, described on AE as diameter, in M; draw AM, meeting the circle BCD in D; and draw EH parallel to AM; join DH; it touches both the given circles.

For MD, EII, which (const.) are parallel, are equal to one another, because each of them is equal to KB; therefore (I. 15, cor. 3) DII is parallel to ME. Now (III. 11), the angle AME in a semicircle is a right angle; and therefore (I. 16) the angles ADII, DIIE are right angles, and (III. 8, cor.) DII touches both the circles, since it is perpendicular to the radii AD, EII.

Scho. In the figure the circles lie on the same side of the tangent, which is therefore exterior to them ; but the tangent can be transverse, or lie between the circles. It is plain also, that in these figures, by using the point L instead of M, another exterior and another transverse tangent would be obtained ; and this will always be so, when each of the circles lies wholly without the other, and does not touch it. But if the circles touch one another externally, the two transverse tangents coalesce into a single line passing through their point of contact ; if they cut one another, there will be two exterior tangents, but no transverse one ; if one of the circles touch the other internally, they can have only one common tangent, and this passes through their point of contact ; and, lastly, if one of them lie wholly within the other without touching it, they can have no common tangent.

If the circles be equal, the points of contact of the exterior common tangents are the extremities of the diameters perpendicular to the line joining the centers ; for (I. 15, cor. 3, and 16) the lines connecting the extremities of these toward the same parts are perpendicular to the diameters, and therefore (III. 8, cor.) they touch the circles.

PROP. XXV.—PROB.—*To inscribe any regular polygon in a given circle.*

Let ABDC be the given circle ; it is required to inscribe any polygon in it.

Since (I. 9, cor.) all the angles formed by any number of straight lines intersecting each other in a common point are equivalent to four right angles, and (I. 20, cor. 1) any rectilinear figure can be divided into as many triangles as the figure has sides, by straight lines joining the extremities of the

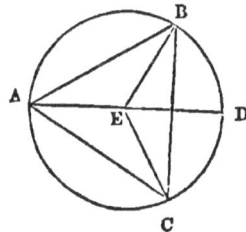

sides with a point within the figure—a regular polygon being equilateral—the straight lines connecting the extremities of those sides with the center of the regular polygon will divide the regular polygon into equal triangles; for if ABC be a regular polygon of three sides, AB is equal BC, and also equal to CA, and draw from E (III. 1, cor. 1), the center of the polygon, EA, EB, and EC. Now the triangles AEB, BEC, and CEA, having their bases equal (hyp.), their other sides common, (const.) are equal; therefore (I. 4) the angles AEB, BEC, and CEA are equal; but those angles are equivalent to form right angles (I. 20, cor. 1); hence, each is one third of four right angles, or two thirds of two right angles; consequently, a regular polygon can be inscribed in a circle by making an angle on the diameter (I. 13) with the center of the circle the vertex of the angle, equal to that part of four right angles that the regular polygon has sides, viz. : if the regular polygon has four sides, each angle at the center will be one fourth of four right angles; if five sides, one fifth of four right angles; and so on. Or, by bisecting (I. 5) the angle or the arc (III. 17), a regular polygon of double the number of sides can be inscribed in the circle. Having by these means the angle at the center of a regular polygon, the chord of the arc intercepted by the sides of the angles will be the side of the regular polygon required, from which (I. 12 and 13) the regular polygon can be inscribed in the given circle, which was to be done.

Cor. 1. If the sides of the angles be produced beyond the circumference (I. post. 2), and parallels to the exterior sides of the polygon (I. 18) be drawn touching the circle (III. 9), then a similar and regular polygon can be described about the given circle.

Cor. 2. Find the center of a given regular polygon (III. 1, cor. 1), and a circle can be inscribed in the polygon, or circumscribed about it, by taking the straight line drawn from the center of the polygon to an extremity of the side of the polygon for a radius for the circumscribed circle, and a line drawn from the center to the point of bisection of the side for a radius of the inscribed circle.

Cor. 3. Since the triangles AEB, BEC, and AEC are equivalent respectively to the rectangles (I. 23, cor. 6) under the radius of an inscribed circle and the halves of AB, BC, and CA, it follows (II. 1) that the area of ABC is equivalent to the rectangle under the radius and half the perimeter. Hence, if the sides be given in numbers, the length of the radius may be computed by calculating (II. 12, scho.) the area, and dividing it by half the sum of the sides, or its double by the sum of the sides.

Cor. 4. Since the angles formed about the center of a circle are together equivalent to four right angles (I. 9, cor.), and since the angles of a triangle are together equivalent to two right angles (I. 20), it follows that when an equiangular triangle is formed (I. 2) having a vertex at the center and the radius for a side, each angle of the triangle is one third of two right angles, or one sixth of four right angles; hence, six equal angles can be formed (I. 13) about the center of a circle, and since the sides about these angles are intercepted by the circumference (const.), they are equal (I. def. 16) ; and (I. 2, cor.) an equiangular triangle being also equilateral, the side of the triangle opposite the angle at the center of the circle is equal to the radius of the circle (I. ax. 1) ; therefore the radius can be made to subtend six equal arcs of the circumference.

Cor. 5. When in the case of a general proposition *to descri'e a circle touching three given straight lines which do not pass through the same point, and which are not all parallel to one another.* If two of the lines be parallel, there may evidently be two equal circles, one on each side of the line falling on the parallels, each of which will touch the three given lines, and their centers will be the intersections of the lines bisecting the angles made by the parallels with the third line. But if the lines form a triangle by their intersections, there will be four circles touching them ; one, inscribed, and the others each touching one side externally and the other two produced. The centers of the external circles will be the intersections of the lines bisecting two exterior angles; and the line bisecting the remote interior angle will pass through the same point. The method of proof is the same as that given in the proposition.

Cor. 6. If straight lines be drawn from the center of one of the external circles to the vertices of the triangle, the three triangles formed by the sides of the triangle and the straight lines are respectively equivalent to the rectangles (L 15, cor. 4) under the radius of that circle and halves of the sides of the triangle. And if the triangle formed by the side of the original triangle nearest the center of the circle, and the lines drawn to the vertices at the extremities of that side, be taken from the sum of the two other triangles, there remains the original triangle equivalent to the rectangle under the radius, and the excess of half the sum of the two other sides of the original triangle above half the side nearest the center of the circle, or, which is the same thing, to the rectangle under the radius, and the excess of half the perimeter of the original triangle above the side nearest the center of the circle. The radius, therefore, of any of the external circles may be computed by dividing the area of the original triangle by the excess of half its perimeter above the side which the circle touches externally.

Scho. The polygons considered in this proposition, and those which may be derived from them by the process of bisecting the angles or arcs subtending the sides of the polygons, are the only ones till lately which geometers have been able to describe by elementary geometry, that is, by means of the straight line and circle. M. Gauss, of Gottingen, in his *Disquisitiones Arithmeticæ*, has shown that by elementary geometry, every regular poly_on may be inscribed in a circle, the number of whose sides is a power of 2 increased by unity, and is a prime number, or a number which can not be produced by the multiplication of two whole numbers, such as 17—the fourth power of 2 increased by unity—and polygons of 257 and 65537 sides. But the investigation is too complex and difficult for an ordinary school text-book.

END OF BOOK THIRD.

BOOK FOURTH.*

THE GENERAL THEORY OF PROPORTION.

DEFINITIONS.

1. A LESS number or magnitude is said to *measure* a greater, or to be a *measure*, a *part*, or a *submultiple* of the greater, when the less is contained a certain number of times, exactly, in the greater; and

2. The greater is said to be a *multiple* of the less.

3. Magnitudes which can be compared in respect of quantity, that is, which are either equal to one another, or unequal, are said to be of the *same kind*, or *homogeneous*.

Scho. 1. Thus, lines, whether straight or curved, are magnitudes of the same kind, or are *homogeneous*, since they may be equal or unequal. In like manner, surfaces, solids, and angles form three other classes of homogeneous magnitudes. On the contrary, lines and surfaces, lines and angles, surfaces and solids, etc., are heterogeneous. Thus, it is obviously improper to say, that the side and area of a square are equal to one another, or are unequal. So likewise we can not say that the area and one of the angles of a triangle are equal to each other, or are unequal; and they are therefore heterogeneous.

4. If there be two magnitudes of the same kind, the relation which one of them bears to the other in respect of quantity, is called its *ratio* to the other.

The first term, or magnitude, is called the *antecedent* of the ratio, and the second the *consequent*.

5. If there be four magnitudes, and if any like multiples whatever be taken of the first and third, and any whatever of

---

* The accurate but prolix method of Euclid is substituted by the following more concise method, by employing the notations and simple principles of Algebra. See *Sup.* to Book V. Thomson's Euclid.

the second and fourth; and if, according as the multiple of the first is greater than the multiple of the second, equal to it, or less, the multiple of the third is also greater than the multiple of the fourth, equal to it, or less; then the first of the magnitudes is said to have to the second the *same ratio* that the third has to the fourth.

6. Magnitudes which have the same ratio are called *proportionals ;* and equality or identity of ratios constitutes *proportion* or *analogy.*

When magnitudes are proportionals, the relation is expressed briefly by saying, that the first *is to* the second, *as* the third *to* the fourth, the fifth *to* the sixth, and so on.

7. When of the multiples of four magnitudes, taken as in the fifth definition, the multiple of the first is greater than that of the second, but the multiple of the third is not greater than that of the fourth; then the first is said to have to the second a *greater ratio* than the third has to the fourth; and, on the contrary, the third is said to have to the fourth a *less ratio* than the first has to the second.

8. When there are three or more magnitudes of the same kind, such that the ratios of the first to the second, of the second to the third, and so on, whatever may be their number, are all equal; the magnitudes are said to be *continual proportionals.*

9. The second of three continual proportionals is said to be a *mean proportional* between the other two.

10. When there is any number of magnitudes of the same kind, greater than two, the first is said to have to the last the ratio compounded of the ratio which the first has to the second, and of the ratio which the second has to the third, and of the ratio which the third has to the fourth, and so on to the last magnitude.

For example, if A, B, C, D be four magnitudes of the same kind, the first, A, is said to have to the last, D, the ratio compounded of the ratios of A to B, B to C, and C to D.

11. When three magnitudes are continual proportionals, the ratio of the first to the third is said to be *duplicate* of the ratio of the first to the second, or of the second to the third.

12. When four magnitudes are continual proportionals, the

ratio of the first to the fourth is said to be *triplicate* of the ratio of the first to the second, of the ratio of the second to the third, or of the ratio of the third to the fourth.

Scho. 2. In continual proportionals, by their own nature, and that of compound ratio, the ratio of the first to the third is compounded of two equal ratios; and the ratio of the first to the fourth, of three equal ratios; and hence we see the reason and the propriety of calling the first duplicate ratio, and the second triplicate. It is plain, that on similar principles, the ratio of the first to the fifth would be said to be *quadruplicate* of the ratio of the first to the second, of the second to the third, etc., and thus we might form other similar terms at pleasure.

The terms *subduplicate*, *subtriplicate*, and *sesquiplicate*, which are sometimes employed by mathematical writers, are easily understood after the explanations given above. In continual proportionals, the ratio of the first term to the second is said to be *subduplicate* of the ratio of the first to the third, and *subtriplicate* of that of the first to the fourth. Again: if there be four continual proportionals, the ratio of the first to the fourth is said to be *sesquiplicate* of the ratio of the first to the third; or, which amounts to the same, the ratio which is compounded of another ratio and its subduplicate, is sesquiplicate of that ratio.

13. In proportionals, the antecedent terms are called *homologous* to one another, as also the consequents to one another.

Geometers make use of the following technical words to denote different modes of deriving one proportion from another, by changing either the order or the magnitudes of the terms.

14. *Alternately:* this word is used when there are four proportionals of the same kind; and it is inferred that the first has the same ratio to the third which the second has to the fourth; or that the first is to the third as the second to the fourth; as is shown in the fourth proposition of this book.

15. *By inversion:* when there are four proportionals, and it is inferred that the second is to the first as the fourth to the third. Prop. 3, Book IV.

16. *By composition:* when there are four proportionals, and it is inferred, that the first, together with the second, is to the

second as the third, together with the fourth is to the fourth. Tenth Prop., Book IV.

17. *By division:* when there are four proportionals, and it is inferred that the excess of the first above the second is to the second, as the excess of the third above the fourth is to the fourth. Tenth Prop., Book IV.

18. *By conversion:* when there are four proportionals, and it is inferred that the first is to its excess above the second, as the third to its excess above the fourth. Eleventh Prop., Book IV.

Scho. The substance of the five preceding definitions may be exhibited briefly in the following manner, the signs $+$ and $-$ denoting addition and subtraction, as has been explained already at the beginning of the second book:

	Let $A : B :: C : D$;
Alternately,	$A : C :: B : D$;
By inversion,	$B : A :: D : C$;
By composition,	$A+B : B :: C+D : D$;
By division,	$A-B : B :: C-D : D$;
By conversion,	$A : A-B :: C : C-D.$

19. *Ex æquo,* or *ex equali* (scil. *distantiâ*), *from equality of distance:* when there is any number of magnitudes more than two, and as many others, which, taken two in the one rank, and two in the other, in *direct* order, have the same ratio; and it is inferred that the first has to the last of the first rank the same ratio which the first of the other rank has to the last. This is demonstrated in the thirteenth proposition of this book.

20. *Ex æquo, inversely:* when there are three or more magnitudes, and as many others, which taken two and two in a *cross* order, have the same ratio; that is, when the first magnitude is to the second in the first rank, as the last but one is to the last in the second rank; and the second to the third of the first rank, as the last but two is to the last but one of the second rank, and so on; and it is inferred, as in the preceding definition, that the first is to the last of the first rank, as the first to the last of the other rank. This is proved in the fourteenth proposition of this book.

AXIOMS.

1. Like multiples of the same, or of equal magnitudes, are equal to one another.

2. Those magnitudes of which the same, or equal magnitudes, are like multiples, are equal to one another.

3. A multiple of a greater magnitude is greater than the same multiple of a less.

4. That magnitude of which a multiple is greater than the same multiple of another, is greater than that other magnitude.

EXPLANATION OF SIGNS.

1. The product arising from multiplying one number by another is expressed by writing the letters representing them, one after the other, without any sign between them; and sometimes by placing between them a point, or the sign $\times$.

2. A product is called a *power*, when the factors are all the same. Thus, AA, or as it is generally written, A^2, is called the second power, or the square of A; AAA, or A^3, the third power, or cube of A; $AAAA$, or A^4, its fourth power, etc.

In relation to these powers, A is called their *root*. Thus, A is the second or square root of A^2, the third or cube root of A^3, the fourth root of A^4, etc. In like manner, the *second* or *square root* of A is a number which, when multiplied by itself, produces A; the *th rd* or *cube ro t* of A is such a number, that if it be multiplied by itself, and the product by the same root again, the final product will be A. The square root of A is denoted by $\sqrt{A}$ or $A^{\frac{1}{2}}$, its cube root by $\sqrt[3]{A}$, or $A^{\frac{1}{3}}$, its fourth root by $A^{\frac{1}{4}}$, etc.

3. The quotient arising from dividing one number by another is denoted by writing the dividend as the numerator of a fraction, and the divisor as its denominator.

4. The signs $=$, $\eqsim$, $>$, $<$, signify respectively *equal to, equivalent to, greater than, less than.*

PROPOSITIONS.

PROP. I.—THEOR.—*If there be four numbers such that the quotients obtained by dividing the first by the second, and the*

7

third by the fourth, are equal; the first has to the second the same ratio that the third has to the fourth.

Let A, B, C, D be four magnitudes, such that $\frac{A}{B}=\frac{C}{D}$; then A: B :: C : D.

For, let m and n be any whole numbers, and multiply the fractions $\frac{A}{B}$ and $\frac{C}{D}$ by m, and divide the product by n; then $\frac{mA}{nB}=\frac{mC}{nD}$. Now, if mA be greater than nB, mC will also be greater than nD; for, if this were not so, $\frac{mA}{nB}$ would not be equal to $\frac{mC}{nD}$. In like manner it might be shown, that if mA be equal to nB, mC will be equal to nD; and that if mA be less than nB, mC will be less than nD. But mA, mC are any like multiples whatever of A, C; and nB, nD any whatever of B, D; and therefore (IV. def. 5) A : B :: C : D. Therefore, if there be four numbers, etc.

Scho. This proposition is the same, when $\frac{A}{B}$ or $\frac{C}{D}=p$ or $\frac{1}{p}$, p being a whole number.

Cor. If AD=BC, by dividing by B and D, we get $\frac{A}{B}=\frac{C}{D}$; and therefore, by this proposition, A : B :: C : D. Hence, if the product of two numbers be equal to that of two others, the one pair may be taken as the extremes and the other as the means of an analogy.

PROP. II.—THEOR.—*If any four numbers be proportional, and if the first be divided by the second, and the third by the fourth, the quotients are equal.*

Let A : B :: C : D; then $\frac{A}{B}=\frac{C}{D}$.

If A and B be whole numbers, let the first and third terms be multiplied by B, and the second and fourth by A, and the products are AB, AB, BC, AD. Now, since the first and second of these are the same, the third and fourth are (IV. def. 5) equal; that is, AD=BC; and by dividing these by B and D, we find (IV. ax. 2) $\frac{A}{B}=\frac{C}{D}$.

If A and B be fractions, let $A=\dfrac{E}{m}$, and $B=\dfrac{F}{n}$, so that $mA=$ E, and $nB=F$; the numerators and denominators E, F, m, n being whole numbers. Then (hyp.) $\dfrac{E}{m}:\dfrac{F}{n}::C:D$. Multiply the first and third of these by mF, and the second and fourth by nE, and the products are EF, EF, mFC, and nED. Now, the first and second of these being the same, the third and fourth (IV. def. 5) are equal; that is, $nED=mFC$, or $mnAD=$ $mnBC$, since $E=mA$, and $F=nB$. Hence, by dividing these by m and n, we get (IV. ax. 2) $AD=BC$; and the rest of the proof is the same as in the first case. Therefore, if any four numbers, etc.

Scho. 1. If either A or B be a whole number, the proof is included in the second part of the demonstration given above. Thus, if A be a whole number, we have simply $E=A$ and $m=$ 1, and everything will proceed as above. The proof would also be readily obtained by substituting for B as before, but retaining A unchanged.

If A and B be incommensurable, such as the numbers expressing the lengths of the diagonal and side of a square, the lengths of the diameter and circumference of a circle, etc., their ratios may be approximated as nearly as we please. Thus the diagonal of a square is to its side, as $\frac{14}{10}:1$, nearly; as $\frac{141}{100}:1$, more nearly; as $\frac{1414}{1000}:1$, still more nearly, etc. Hence, in such cases we can have no hesitation in admitting the truth of the proposition, as we see that it holds with respect to numbers the ratio of which differs from that of the proposed numbers by a quantity which may be rendered as small as we please—smaller, in fact, than anything that can be assigned.

Scho. 2. This proposition is the same, when $\dfrac{A}{B}=p$ or $\dfrac{1}{p}$, p being a whole number.

Scho. 3. From this proposition and the foregoing, it appears, that if two fractions be equal, the numerator of the one is to its denominator as the numerator of the other to its denominator; and that if the first and second of four proportional numbers be made the numerator and denominator of one fraction, and the third and fourth those of another, the two fractions are equal.

This is the same in substance as that the two expressions, A :
B :: C : D, and $\frac{A}{B}=\frac{C}{D}$, are equivalent, and may be used for
one another.

Cor. 1. It appears in the demonstration of this proposition,
that $AD=BC$; that is, if four numbers be proportionals, the
product of the extremes is equal to the product of the means.
Hence, if the product of the means be divided by one of the extremes, the quotient is the other; and thus we have a proof of
the ordinary arithmetical rule for finding a fourth proportional
to three given numbers.

Cor. 2. It is evident, that if A be greater than B, C must be
greater than D; if equal, equal; and if less, less; as otherwise
$\frac{A}{B}$ and $\frac{C}{D}$ could not be equal.

Cor. 3. If A : B :: C : D, and consequently $\frac{A}{B}=\frac{C}{D}$, by multiplying these fractions by $\frac{m}{n}$, we get $\frac{mA}{nB}=\frac{mC}{nD}$, or $mA : nB ::$
$mC : nD$.

Cor. 4. If A be greater than B, the fraction $\frac{A}{C}$ is evidently
greater than $\frac{B}{C}$, and the fraction $\frac{C}{A}$ less than $\frac{C}{B}$; that is, of two
unequal numbers, the greater has a greater ratio to a third than
the less has; and a third number has a greater ratio to the
less than it has to the greater.

Cor. 5. Conversely, if $\frac{A}{C}$ be greater than $\frac{B}{C}$, A is greater than
B; and, if $\frac{C}{A}$ be less than $\frac{C}{B}$, A is also greater than B.

PROP. III.—THEOR.—*If four numbers be proportionals, they
are proportionals also when taken inversely.*

If A : B :: C : D; then, inversely, B : A :: D : C.

For (IV. 2, cor. 1) $BC=AD$; and hence by dividing by A
and C, we obtain $\frac{B}{A}=\frac{D}{C}$, or (IV. 2, scho. 2) B : A :: D : C.

Therefore, if four numbers, etc.

Prop. IV.—Theor.—*If four num'ers be proportionals, they are also proportionals when taken alternately.*

If $A : B :: C : D$; then, alternately, $A : C :: B : D$.

For (IV. 2, cor. 1) $AD = BC$; whence, by dividing by C and D we get $\dfrac{A}{C} = \dfrac{B}{D}$; or (IV. 2, scho. 2) $A : C :: B : D$. Therefore, if four numbers, etc.

Scho. When the first and second terms are not of the same kind as the third and fourth, the terms can not be taken alternately, as ratios would thus be instituted between heterogeneous magnitudes.

Prop. V.—Theor.—*Equal numbers have the same ratio to the same number; and the same has the same ratio to equal numbers.*

Let A and B be equal numbers, and C a third; then $A : C :: B : C$, and $C : A :: C : B$.

For, A and B being equal, the fractions $\dfrac{A}{C}$ and $\dfrac{B}{C}$ are also equal, or, which is the same, $A : C :: B : C$; and, by inversion, (IV. 3) $C : A :: C : B$. Therefore, equal numbers, etc.

Prop. VI.—Theor.—*Numbers which have the same ratio to the same number are equal; and those to which the same has the same ratio are equal.*

If $A : C :: B : C$, or if $C : A :: C : B$, A is equal to B.

For, since $\dfrac{A}{C} = \dfrac{B}{C}$, by multiplying by C we get $A = B$.

The proof of the second part is the same as this, since, by inversion (IV. 3), the second analogy becomes the same as the first. Therefore, numbers, etc.

Prop. VII.—Theor.—*Ratios that are equal to the same ratio are equal to one another.*

If $A : B :: C : D$, and $E : F :: C : D$; then $A : B :: E : F$.

For, since $\dfrac{A}{B} = \dfrac{C}{D}$, and $\dfrac{E}{F} = \dfrac{C}{D}$, therefore (I. ax. 1) $\dfrac{A}{B} = \dfrac{E}{F}$; that is, (IV. 2, scho. 2) $A : B :: E : F$. Therefore, ratios, etc.

PROP. VIII.—THEOR.—*Of numbers which are proportionals, as any one of the antecedents is to its consequent, so are all the antecedents taken together to all the consequents.*

If $A : B :: C : D :: E : F$; then $A : B :: A+C+E : B+D+F$.

Since $\frac{A}{B}=\frac{C}{D}=\frac{E}{F}$, put each fraction equal to q, and multiply by the denominators; then $A=Bq$, $C=Dq$, and $E=Fq$. Hence, by addition, $A+C+E=(B+D+F)q$; and by dividing by $B+D+F$, we get $q=\frac{A+C+E}{B+D+F}$. But $q=\frac{A}{B}$; and therefore $\frac{A}{B}=\frac{A+C+E}{B+D+F}$, or $A : B :: A+C+E : B+D+F$. Therefore, etc.

PROP. IX.—THEOR.—*Magnitudes have the same ratio to one another that their like multiples have.*

Let A and B be two magnitudes; then, n being a whole number, $A : B :: nA : nB$.

For $\frac{A}{B}=\frac{nA}{nB}$, or $A : B :: nA : nB$. Therefore, magnitudes, etc.

PROP. X.—THEOR.—*If four numbers be proportionals; then* (1) *by composition, the sum of the first and second is to the second, as the sum of the third and fourth to the fourth; and* (2), *by division, the excess of the first above the second is to the second, as the excess of the third above the fourth is to the fourth.*

If $A : B :: C : D$; then, by composition, $A+B : B :: C+D : D$; and by division, $A-B : B :: C-D : D$.

1. Since (hyp.) $\frac{A}{B}=\frac{C}{D}$, and since $\frac{B}{B}=\frac{D}{D}$; add the latter fractions to the former, each to each, and there results $\frac{A+B}{B}=\frac{C+D}{D}$, or $A+B : B :: C+D : D$.

2. By subtracting the latter pair of the same fractions from the former, each from each, we obtain $\frac{A-B}{B}=\frac{C-D}{D}$; or $A-B : B :: C-D : D$. If, therefore, etc.

Cor. By dividing the fractions which were found above by addition, by those which were found by subtraction, we get $\frac{A+B}{A-B}=\frac{C+D}{C-D}$; or (IV. 2, scho. 2) $A+B : A-B :: C+D : C-D$; that is, if four numbers be proportional, the sum of the first and second terms is to their difference, as the sum of the third and fourth terms is to their difference. It is evident, that if B be greater than A, the analogy would become $B+A : B-A :: D+C : D-C$.

PROP. XI.—THEOR.—*If four numbers be proportional; then, by conversion, the first is to its excess above the second, as the third to its excess above the fourth.*

If $A : B :: C : D$; then, by conversion, $A : A-B :: C : C-D$.

For, since (hyp. and inver.) $\frac{B}{A}=\frac{D}{C}$, and since $\frac{A}{A}=\frac{C}{C}$; take the former fractions from the latter, each from each, and there remains $\frac{A-B}{A}=\frac{C-D}{C}$, or (by inver.) $A : A-B :: C : C-D$. Therefore, if four numbers, etc.

PROP. XII.—THEOR.—*If there be numbers forming two or more analogies which have common consequents, the sum of all the first antecedents is to their common consequent, as the sum of all the other antecedents is to their common consequent.*

If $A : B :: C : D$, and $E : B :: F : D$; then $A+E : B :: C+F : D$.

For (hyp.) $\frac{A}{B}=\frac{C}{D}$, and $\frac{E}{B}=\frac{F}{D}$; and hence, by addition, $\frac{A+E}{B}=\frac{C+F}{D}$, or $A+E : B :: C+F : D$. If, therefore, etc.

PROP. XIII.—THEOR.—*If there be three or more numbers, and as many others, which, taken two and two in order, have the same ratio ; then, ex æquo, the first has to the last of the first rank the same ratio that the first has to the last of the second rank.*

If the two ranks of numbers, A, B, C, D, and E, F, G, H, be

such that $A : B :: E : F$, $B : C :: F : G$, and $C : D :: G : H$; then $A : D :: E : H$.

For, since (hyp.) $\frac{A}{B} = \frac{E}{F}$, $\frac{B}{C} = \frac{F}{G}$, and $\frac{C}{D} = \frac{G}{H}$; by multiplying together the first, third, and fifth fractions, and the second, fourth, and sixth, we obtain $\frac{ABC}{BCD} = \frac{EFG}{FGH}$; or, by dividing the terms of the first of these fractions by BC, and those of the second by FG, $\frac{A}{D} = \frac{E}{H}$, or $A : D :: E : H$. Therefore, if there be three, etc.

This proposition might also be enunciated thus: If there be numbers forming two or more analogies, such that the consequents in each are the antecedents in the one immediately following it, an analogy will be obtained by taking the antecedents of the first analogy and the consequents to the last for its antecedents and consequents.

PROP. XIV.—THEOR.—*If there be three or more numbers, and as many others, which, taken two and two in a cross order, have the same ratio ; then,* ex æquo inversely, *the first has to the last of the first rank the same ratio which the first has to the last of the second rank.*

If the two ranks of numbers, A, B, C, D, and E, F, G, H, be such that $A : B :: G : H$, $B : C :: F : G$, and $C : D :: E : F$; then, *ex æquo inversely*, $A : D :: E : H$.

For, since (hyp.) $\frac{A}{B} = \frac{G}{H}$, $\frac{B}{C} = \frac{F}{G}$, and $\frac{C}{D} = \frac{E}{F}$, by multiplying together the fractions as in the preceding proposition, we get $\frac{ABC}{BCD} = \frac{GFE}{HGF}$; whence, by dividing the terms of the first of these fractions by BC, and those of the second by GF, we obtain $\frac{A}{D} = \frac{E}{H}$, or $A : D :: E : H$. If, therefore, etc.

This proposition may also be enunciated thus: If there be numbers forming two or more analogies, such that the means of each are the extremes of the one immediately following it, another analogy may be obtained by taking the extremes of the first analogy and the means of the last for its extremes and means.

PROP. XV.—THEOR.—*If there be numbers forming two or more analogies, the products of their corresponding terms are proportionals.*

If $A : B :: C : D$, $E : F :: G : H$, and $K : L :: M : N$; then $AEK : BFL :: CGM : DHN$.

For (hyp.) $\dfrac{A}{B} = \dfrac{C}{D}$, $\dfrac{E}{F} = \dfrac{G}{H}$, and $\dfrac{K}{L} = \dfrac{M}{N}$; and taking the products of the corresponding terms of these fractions, we obtain $\dfrac{AEK}{BFL} = \dfrac{CGM}{DHN}$, or $AEK : BFL :: CGM : DHN$. Therefore, if there be numbers, etc.

Cor. 1. Hence, if there be two analogies consisting of the same terms, A, B, C, D, we have $A^2 : B^2 :: C^2 : D^2$; if there be three, we have $A^3 : B^3 :: C^3 : D^3$, etc.; and it thus appears, that like powers of proportional numbers are themselves proportional.

Cor. 2. Like roots of proportional numbers are proportional. Thus, if $A : B :: C : D$, let $\sqrt{A} : \sqrt{B} :: \sqrt{C} : \sqrt{E}$. Then, by the preceding corollary, $A : B :: C : E$. But (hyp.) $A : B :: C : D$; and therefore (IV. 7) $C : E :: C : D$, and (IV. 6) $E = D$, and consequently $\sqrt{} : A \sqrt{} B :: \sqrt{C} : \sqrt{E}$, or $\sqrt{D}$.

PROB. XVI.—THEOR.—*The sum of the greatest and least of four proportional numbers is greater than the sum of the other two.*

If $A : B :: C : D$, and if A be the greatest, and therefore (IV. 2, cor. 2) D the least; A and D are together greater than B and C.

For (by conversion) $A : A—B :: C : C—D$, and, alternately, $A : C :: A—B : C—D$. But (hyp.) $A > C$, and therefore (IV. 2, cor. 2) $A—B > C—D$. To each of these add B; then $A > B + C—D$. Add again, D; then, $A + D > B + C$. Therefore, etc.

Cor. Hence the mean of three proportional numbers is less than half the sum of the extremes.

PROB. XVII.—THEOR.—*In numbers which are continual proportionals, the first is to the third as the second power of the first to the second power of the second; the first to the fourth as the third power of the first to the third power of the second; the first to the fifth as the fourth power of the first to the fourth power of the second; and so on.*

If A, B, C, D, E, etc., be continual proportionals; A : C :: A^2 : B^2; A : D :: A^3 : B^3; A : E :: A^4 : B^4, etc.

For, since (IV. def. 8) A : B :: B : C, and since A : B :: A : B, we have (IV. 15) A^2 : B^2 :: AB : BC, or, dividing the third and fourth terms by B, A^2 : B^2 :: A : C.

Again: since A^2 : B^2 :: A : C,

and A : B :: C : D we have (IV. 15) A^3 : B^3 :: AC : CD, or dividing the third and fourth terms by C, A^3 : B^3 :: A : D; and so on, as far as we please. Therefore, etc.

Cor. Hence (IV. defs. 11 and 12) the ratio which is duplicate of that of any two numbers, is the same as the ratio of their squares; that which is triplicate of their ratio, the same as the ratio of their cubes, etc.

PROP. XVIII.—THEOR.—*A ratio which is compounded of other ratios, is the same as the ratio of the products of their homologous terms.*

Let the ratio of A to D be compounded of the ratios of A to B, B to C, and C to D; the ratio of A to D is the same as that of ABC, the product of the antecedents, to BCD, the product of the consequents.

For, since A : D :: A : D, multiply the terms of the second ratio by BC; then (IV. 9) A : D :: ABC : BCD. Therefore, etc.

PROP. XIX.—THEOR.—*In numbers which are continual proportionals, the difference of the first and second is to the first, as the difference of the first and last is to the sum of all the terms except the last.*

If A, B, C, D, E be continual proportionals, A—B : A :: A—E : A+B+C+D.

For, since (hyp.) A : B :: B : C :: C : D :: D : E, we have (IV. 8) A : B :: A+B+C+D : B+C+D+E. Hence (conv.) A : A—B :: A+B+C+D : A—E; and (inver.) A—B : A :: A—E : A+B+C+D.

It is evident that if A were the least term, and E the great-est, we should get in a similar manner, $B - A : A :: E - A : A + B + C + D$. Therefore, in numbers, etc.

Cor. If the series be an infinite decreasing one, the last term will vanish, and if S be put to denote the sum of the series, the analogy will become $A - B : A :: A : S$; and this, if rA be put instead of B, and the first and second terms be divided by A, will be changed into $1 - r : 1 :: A : S$. The number r is called the *common ratio*, or *common multiplier*, of the series, as by multiplying any term by it, the succeeding one is obtained.

END OF BOOK FOURTH.

BOOK FIFTH.

1. *Similar rectilineal figures* are those which have their several angles equal, each to each, and the sides about the equal angles proportionals.

2. Two magnitudes are said to be *reciprocally proportional* to two others, when one of the first pair is to one of the second, as the remaining one of the second is to the remaining one of the first.

3. A straight line is said to be cut in *extreme and mean ratio*, when the whole is to one of the segments as that segment is to the other.

4. The *altitude* of any figure is the straight line drawn from its vertex perpendicular to its base.

5. A straight line is said to be cut *harmonically*, when it is divided into three segments, such that the whole line is to one of the extreme segments as the other extreme segment is to the middle one.

PROPOSITIONS.

Prop. I.—Theor.—*Triangles and parallelograms of the same altitude are one to another as their bases.*

Let the triangles ABC, ACD, and the parallelograms EC, CF have the same altitude, viz., the perpendicular drawn from the point A to BD; then, as the base BC is to the base CD, so is the triangle ABC to the triangle ACD, and the parallelogram EC to the parallelogram CF.

Produce BD both ways, and take any number of straight lines BG, GH, each equal to BC; and any number DK, KL, each equal to CD; and join AG, AH, AK, AL. Then, because CB, BG, GH are all equal, the triangles ABC, AGB, AHG are (L 15, cor.) all equal. Therefore, whatever multiple the base

HC is of BC, the same multiple is the triangle AHC of ABC.
For the same reason, whatver multiple LC is of CD, the same
multiple is the triangle ALC
of ADC. Also, if the base HC
be equal to CL, the triangle
AHC is equal (I. 15, cor.) to
ALC; and if the base HC be
greater than CL, likewise (I.
15, cor. 6) the triangle AHC is
greater than ALC; and if less, less. Therefore, since there are
four magnitudes, viz., the two bases, BC, CD, and the two triangles ABC, ACD; and of the base BC, and the triangle ABC,
the first and third, any like multiples whatever have been taken,
viz., the base HC, and the triangle AHC; and of the base CD,
and the triangle ACD, the second and fourth, have been taken
any like multiples whatever, viz., the base CL, and the triangle
ALC; and that it has been shown that, if the base HC be
greater than CL, the triangle AHC is greater than ALC; if
equal, equal; and if less, less; therefore (IV. def. 5) as the
base BC is to the base CD, so is the triangle ABC to the triangle ACD.

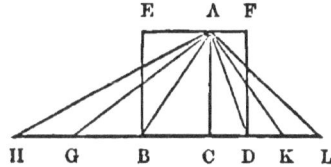

Again: because (I. 15, cor.) the parallelogram CE is double
of the triangle ABC, and the parallelogram CF of the triangle
ACD, and that (IV. 9) magnitudes have the same ratio which
their like multiples have; as the triangle ABC is to the triangle ACD, so is the parallelogram EC to the parallelogram CF.
But it has been shown, that BC is to CD, as the triangle ABC
to the triangle ACD; and as the triangle ABC is to the triangle ACD, so is the parallelogram EC to the parallelogram CF;
therefore (IV. 7) as the base BC is to the base CD, so is the
parallelogram EC to the parallelogram CF. Wherefore, triangles, etc.

Scho. This proposition may be briefly demonstrated thus:
Let a perpendicular drawn from A to BD be called P. Then,
$\frac{1}{2}$P.BC will be equivalent to the area of the triangle ABC, and
$\frac{1}{2}$P.CD that of ACD. Dividing, therefore, the former of these
equals by the latter, we get $\frac{\frac{1}{2}P.BC}{\frac{1}{2}P.CD}$ or, $\frac{BC}{CD} = \frac{ABC}{ACD}$, or (IV. 2,
scho. 2) BC : CD :: ABC : ACD. In extending this method

of proof to the parallelograms, we have merely to use P instead of $\frac{1}{4}$P.

Cor. 1. From this it is plain, that triangles and parallelograms which have equal altitudes, are one to another as their bases.

Let the figures be placed so as to have their bases in the same straight line; and perpendiculars being drawn from the vertices of the triangles to the bases, the straight line which joins the vertices is parallel (I. 15, cor.) to that in which their bases are, because the perpendiculars are both equal and parallel to one another. Then, if the same construction be made as in the proposition, the demonstration will be the same.

Cor. 2. Hence, if A, B, C be any three straight lines, we have A : B :: A.C : B.C.

Cor. 3. So, likewise, if the straight lines A, B, C, D be proportional, and E and F be any other straight lines, we shall have, according to the preceding corollary, and the seventh proposition of the fourth book, A.E : BE :: C.F : D.F.

Prop. II.—Theor.—*If a straight line be parallel to the base of a triangle, it cuts the other sides, or those produced, proportionally, and the segments between the parallel and the base are homologous to one another; and* (2) *if the sides of a triangle, or the sides produced, be cut proportionally, so that the segments between the points of section and the base are homologous to one another, the straight line which joins the points of section is parallel to the base.*

The enunciation of this proposition which is given by Dr. Simson and others, is defective, and might lead to error in its application, as it does not point out what lines are homologous to one another in the analogies.

It is plain that, instead of one proposition, this is in reality two, which are converses of one another.

1. Let DE be parallel to BC, one of the sides of the triangle ABC ; BD : DA :: CE : EA.

Join BE, CD. Then (I. 15, cor.) the triangles BDE, CDE are equivalent, because they are on the same base DE, and between the same parallels. DE, BC. Now ADE is another triangle, and (IV. 5) equal magnitudes have to the same the same ratio; therefore, as the triangle BDE to ADE, so is the triangle CDE to ADE. But,

(V. 1) as the triangle BDE to ADE, so is BD to DA;

Because, having the same altitude, viz., the perpendicular drawn from E to AB, they are to one another as their bases; and for the same reason,

as the triangle CDE to ADE, so is CE to EA.

Therefore (IV. 7) as BD : DA :: CE : EA.

2. Next, let the sides AB, AC of the triangle ABC, or those produced, be cut proportionally in the points D, E; that is, so that BD : DA :: CE : EA, and join DE; DE is parallel to BC.

The same construction being made, because (hyp.)

as BD : DA :: CE : EA; and (V. 1)

as BD to DA, so is the triangle BDE to the triangle ADE; and as CE to EA, so is the triangle CDE to ADE; therefore (IV. 7) the triangle BDE is to ADE, as the triangle CDE to ADE; that is, the triangles BDE, CDE have the same ratio to ADE; and therefore (IV. 6) the triangles BDE, CDE are equal; and they are on the same base DE, and on the same side of it; therefore (I. 15, cor.) DE is parallel to BC. Wherefore, if a straight line, etc.

Cor. The triangles which two intersecting straight lines form with two parallel ones, have their sides which are on the intersecting lines proportional; and those sides are homologous which are in the same straight line; and (2), conversely, if two straight lines form with two intersecting ones triangles which have their sides that are on the intersecting lines proportional, the sides which are in the same straight line with one another being homologous, those straight lines are parallel.

1. Let DE and BC (first and second figures) be the parallels, and let them be cut by the straight lines BD, CE, which intersect each other in A; then BA : AC :: DA : AE. For, since BD : DA :: CE : EA, we have, by composition in the first fig-

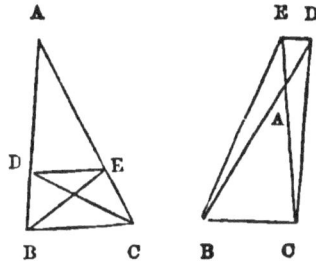

ure, and by division in the second, BA : DA :: CA : EA, and,
alternately, BA : AC :: DA : AE.

2. But if BA : AC :: DA : AE, DE and BC are parallel.
For, alternately, BA : DA :: CA : EA ; then, in the first figure
by division, and in the second by composition, we have BD :
DA :: CE : EA ; and therefore, by the second part of this
proposition, DE is parallel to BC.

PROP. III.—THEOR.— *The sides about the equal angles of
equiangular triangles are proportionals ; and those which are
opposite to the equal angles are homologous sides, that is, are
the antecedents or consequents of the ratios.*

Let ABC, DCE be equiangular triangles, having the angle
ABC equal to DCE, and ACB to DEC, and consequently (I.
20, cor. 5) BAC equal to CDE ; the sides about the equal
angles are proportionals ; and those are the homologous sides
which are opposite to the equal angles.

Let the triangles be placed on the same side of a straight
line BE, so that sides BC, CE, which are opposite to equal
angles, may be in that straight line and contiguous to one another ; and so that neither the equal angles ABC, DCE, nor
ACB, DEC at the extremities of those sides may be adjacent.

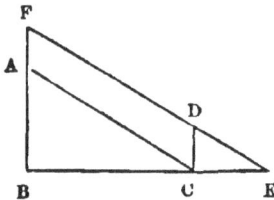

Then, because (I. 20) the angles
ABC, ACB are together less than
two right angles, ABC and DEC,
which (hyp.) is equal to ACB, are
also less than two right angles ;
wherefore (I. 19) BA, ED will meet,
if produced ; let them be produced
and meet in F. Again: because
the angle ABC is equal to DCE, BF is parallel (I. 16, cor.) to
CD ; and, because the angle ACB is equal to DEC, AC is parallel to FE. Therefore, FACD is (I. def. 15) a parallelogram ;
and consequently (I. 15, cor.) AF is equal to CD, and AC to
FD. Now (V. 2) because AC is parallel to FE, one of the
sides of the triangle FBE,

BA : AF :: BC : CE.

But AF is equal to CD ; therefore (IV. 5)

as BA : CD :: BC : CE,

and alternately (IV. 4) as AB : BC :: DC : CE.
Again: (V. 2) because CD is parallel to BF, as

BC : CE :: FD : DE; but FD is equal to AC; therefore,

as BC : CE :: AC : DE; and, alternately,

as BC : CA :: CE : ED.

Therefore, because it has been proved that

AB : BC :: DC : CE, and as BC : CA :: CE : ED;

ex æquo (IV. 13), BA : AC :: CD : DE. Therefore, the sides, etc.

Scho. 1. Hence (V. def. 1) equiangular triangles are similar.

Cor. If two angles of one triangle be respectively equal to two angles of another, their sides are proportional, and the sides opposite to equal angles are homologous. For (I. 20, cor. 5) the remaining angles are equal, and therefore the triangles are equiangular.

Scho. 2. In a similar manner we may produce a given straight line, so that the whole line so produced may have to the part produced the ratio of two given straight lines. Thus, if BA be the line to be produced, make at B an angle of any magnitude, and take BE and CE equal to the other given lines; join AC, and draw FE parallel to it. Then, since FE is parallel to AC, a side of the triangle ABC, we have (V. 2) BF : AF :: BE : CE, so that BF has to AF the given ratio.

PROP. IV.—THEOR.—*The straight line which bisects an angle of a triangle, divides the opposite side into segments which have the same ratio to one another as the adjacent sides of the triangle have; and (2) if the segments of the base have the same ratio as the adjacent sides, the straight line drawn from the vertex to the point of section, bisects the vertical angle.*

1. Let the angle BAC of the triangle ABC be bisected by the straight line AD; then BD : DC :: BA : AC.

Through C draw (I. 18) CE parallel to DA; then (I. 16, cor. 1)
BA produced will meet CE; let
them meet in E. Because AC
meets the parallels AD, EC, the
angle ACE is equal (I. 16) to the
alternate angle CAD; and because BAE meets the same paral-

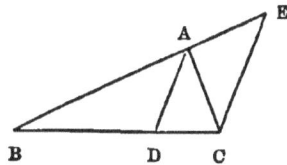

8

lels, the angle E is equal (I. 16, part 2) to BAD; therefore (I. ax. 1) the angles ACE, AEC are equal, because they are respectively equal to the equal angles, DAC, DAB; and consequently AE is equal (I. 1, cor.) to AC. Now (V. 2) because AD is parallel to EC, one of the sides of the triangle BCE, BD : DC :: BA : AE; but AE is equal to AC; therefore (IV. 5) BD : DC :: BA : AC.

2. Let now BD : DC :: BA : AC, and join AD; the angle BAC is bisected by AD.

The same construction being made, because

(hyp.) BD : DC :: BA : AC; and

(V. 2) BD : DC :: BA : AE,

since AD is parallel to EC; therefore (IV. 7) BA : AC :: BA : AE; consequently (IV. 6) AC is equal to AE; and (I. 1) the angles AEC, ACE are therefore equal. But (I. 16) the angle BAD is equal to E, and DAC to ACE; wherefore, also, BAD is equal (I. ax. 1) to DAC; and therefore the angle BAC is bisected by AD. The straight line, therefore, etc.

And if an exterior angle of a triangle be bisected by a straight line which also cuts the base produced, the segments between the bisecting line and the extremities of the base have the same ratio to one another as the other sides of the triangle have; and (2) if the segments of the base produced have the same ratio which the other sides of the triangle have, the straight line drawn from the vertex to the point of section bisects the exterior angle of the triangle.

1. Let an exterior angle CAE of any triangle ABC be bisected by AD which meets the opposite side produced in D; then BD : DC :: BA : AC.

Through C draw (I. 18) CF parallel to AD; and because AC meets the parallels AD, FC, the angle ACF is equal (I. 16) to CAD; and because the straight line FAE meets the parallels AD, FC, the angle CFA is equal to DAE; therefore, also, ACF, CFA are (I. ax. 1) equal to one another, because they are respectively equal to the equal angles DAC, DAE; and consequently (I. cor.) AF is equal to AC. Then (V. 2) because AD is parallel to FC, a side

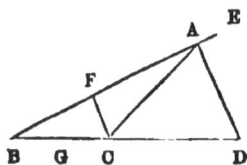

of the triangle BCF, BD : DC :: BA : AF; but AF is equal to AC; as therefore BD : DC :: BA : AC.

2. Let now BD : DC :: BA : AC, and join AD; the angle CAD is equal to DAE.

The same construction being made, because
BD : DC :: BA : AC; and that (V. 2) BD : DC :: BA : AF; therefore (IV. 7) BA : AC :: BA : AF; wherefore (IV. 6) AC is equal to AF, and (I. 1) the angle AFC to ACF. But (I. 16) the angle AFC is equal to EAD, and ACF to CAD; therefore, also (I. ax. 1), EAD is equal to CAD. Wherefore, etc.

Cor. If G be the point in which BC is cut by the straight line bisecting the angle BAC,

we have (V. 4) BG : GC :: BA : AC;

and by this proposition, BD : DC :: BA : AC;

whence (IV. 7) BD : DC :: BG : GC, and therefore (V. def. 5) BD is divided harmonically in G and C.

Scho. If the triangles be isosceles, the line bisecting the exterior angle at the vertex is parallel to the base. In this case, the segments may be regarded as infinite, and therefore equal, their difference, the base, being infinitely small in comparison of them.

PROP. V.—THEOR.—*If the sides of two triangles, about each of their angles, be proportionals, the triangles are equiangular, and have their equal angles opposite to the homologous sides.*

Let the triangles ABC, DEF have their sides proportionals, that AB : BC :: DE : EF; and BC : CA :: EF : FD; and consequently, *ex æquo*, BA : AC :: ED : DF; the triangles are equiangular, and the equal angles are opposite to the homologous sides, viz., the angle ABC equal to DEF, BCA to EFD, and BAC to EDF.

At the points E, F, in the straight line EF, make (I. 13) the angle FEG equal to B, and EFG equal to C. Then (V. 3, cor.) the triangles ABC, GEF have their sides opposite to the equal angles proportionals ; wherefore,

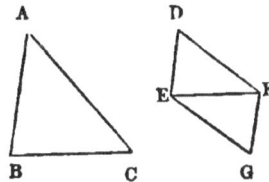

AB : BC :: GF : EF; but (hyp.)
AB : BC :: DE : EF.

Therefore (IV. 7) DE : EF :: GF : EF; whence, since DE and GF have the same ratio to EF, they are (IV. 6) equal. It may be shown in a similar manner that DF is equal to EG; and because, in the triangles DEF, GEF, DE is equal to FG, EF common, and DF equal to GE; therefore (I. 4) the angle DEF is equal to GFE, DFE to GEF, and EDF to EGF. Then, because the angle DEF is equal to GFE, and (const.) GFE to ABC; therefore the angle ABC is equal to DEF. For the same reason, ACB is equal to DFE, and A to D. Therefore the triangles ABC, DEF are equiangular. Wherefore, if the sides, etc.

PROP. VI.—THEOR.—*If two triangles have one angle of the one equal to one angle of the other, and the sides about the equal angles proportionals ; the remaining angles are equal, each to each, viz., those which are opposite to the homologous sides.*

Let the triangles ABC, GEF, of the previous diagrams, have the angles ABC, EFG equal, and the sides about those angles proportionals; that is, BA : BC :: GF : EF ; the angle BAC is equal to EGF, and ACB to EFG.

Make (I. 13) the angle FED equal to either of the angles ABC, GFE; and the angle EFD equal to ACB. Then (V. 3, cor.),

BA : BC :: DE : EF. But (hyp.)
BA : BC :: GF : EF ;

and therefore (IV. 7) GE : EF :: DE : EF; wherefore ED is equal (IV. 6) to FG. Now EF is common to the two triangles GEF, DEF; and the angle GFE is equal (const.) to DEF; therefore the angle EFD is equal (I. 3) to FEG, and D to G. But (const.) the angle EFD is equal to ACB; therefore ACB is equal to FEG; and (hyp.) the angle ABC is equal to GEF; wherefore, also (I. 20, cor 5), the remaining angles A and G are equal. Therefore, if two triangles, etc.

PROP. VII.—THEOR.—*If two triangles have two sides of the one proportional to two sides of the other, and if the angles opposite to one pair of the homologous sides be equal, and those opposite to the other pair be either both acute, or not acute, the angles contained by the proportional sides are equal.*

Let ABC and EFG be two triangles which have the sides CB, CA proportional to GF, GE; the angles CBA, GFE equal, and the other angles CAB, GEF acute; then the angles ACB, EGF, contained by the proportional sides, are equal.

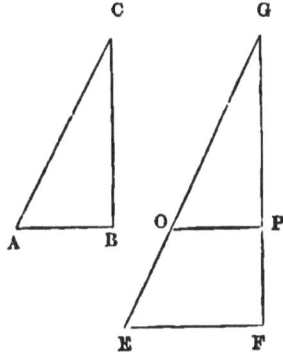

If the triangle ABC be applied to EFG, so that CB will fall on GF, and the vertex C on the vertex G, and make GP equal to CB; then, because the angle CBA is equal to the angle GFE, AB will take the direction OP parallel to EF (I. 16). Since OP is parallel to EF, GOP is equal to GEF (I. 16, cor.), and we have GF : GP :: GE : GO (V. 3); but by hypothesis GF : CB :: GE : CA, and GP is equal to CB. Hence, CA is equal to GO; therefore OP, which joins the extremities of GP and GO, is equal to AB, which joins the extremities of CB and CA (I. ax. 1), and the triangles ABC, OPG are equal. Hence, the angles GOP, CAB are equal (I. 16, cor. 1), but the angle GOP is equal to the angle GEF (I. 16, cor. 1); consequently (I. 14), the remaining angle of GOP is equal to the remaining angle of EFG, and the angles ACB, EGF are equal. In the same manner it can be shown that the angles ACB, EGF are equal when CAB, GEF are not acute. Wherefore, if two triangles have, etc.

Prop. VIII.—Theor.—*In a right-angled triangle, if a perpendicular be drawn from the right angle to the hypothenuse, the triangles on each side of it are similar to the whole triangle, and to one another.*

Let ABC be a right-angled triangle, having the right angle BAC; and from the point A let AD be drawn perpendicular to the hypothenuse BC; the triangles ADB, ADC are similar to the whole triangle ABC, and to one another.

Because the angle BAC is equal (I. ax. 11) to ADB, each of

them being a right angle, and that the angle B is common to
the two triangles ABC, ABD; the remaining angle C is equal

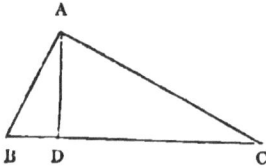

(I. 20, cor. 5) to the remaining angle
BAD. Therefore the triangles ABC,
ABD are equiangular, and (V. 3) the
sides about their equal angles are
proportionals; wherefore (V. def. 1)
the triangles are similar. In the
same manner it might be demon-
strated, that the triangle ADC is equiangular and similar to
ABC; and the triangles ADB, ADC, being each equiangular
to ABC, are (I. ax. 1) equiangular, and therefore (V. 3 and
def. 1) similar to each other. Therefore, etc.

Cor. From this it is manifest, that the perpendicular drawn
from the right angle of a right-angled triangle to the hypothenuse, is a mean proportional (IV. def. 9) between the segments
of the hypothenuse; and also that each of the sides is a mean
proportional between the hypothenuse and its segment adjacent
to that side. For (V. 3) in the triangles BDA, ADC,

as BD : DA :: DA : DC; in the triangles ABC, DBA,
as BC : BA :: BA : BD; and, in the triangles ABC, ACD,
as BC : CA :: CA : CD.

Scho. This proposition affords an easy way of solving the
first corollary to the twenty-fourth proposition of the first book,
as follows:

Let ABC be a triangle, right-angled at A; the square of the
hypothenuse BC is equivalent to the squares of the legs AB,
AC.

Draw AD perpendicular to BC. Then (V. 8, cor.) BC :
BA :: BA : BD, and BC : CA :: CA : CD. Hence (IV. 2,
cor. 1) the rectangle BC.BD is equivalent to the square of AB,
and the rectangle BC.CD to the square of AC. Hence (I.
ax. 2) BC.BD+BC.CD, or (II. 2) $BC^2 \Longleftrightarrow AB^2 + AC^2$.

PROP. IX.—PROB.—*To find a third proportional to two
given straight lines.*

Let A and B be two given straight lines; it is required to
find a third proportional to them.

Take two straight lines CF, CG, containing any angle C; and upon these make CD equal to A, and DF, CE each equal to B. Join DE, and (I. 18) draw FG parallel to it. EG is the third proportional required.

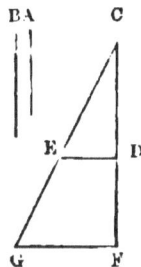

For (V. 2) since DE is parallel to FG, CD : DF :: CE : EG. But (const.) CD is equal to A, and DF, CE each equal to B; therefore A : B :: B : EG; wherefore to A and B the third proportional EG is found, which was to be done.

Scho. Other modes of solving this problem may sometimes be employed with advantage. The following are among the most useful :

1. Draw AD (fig. to prop. 8) perpendicular to the indefinite straight line BC, and make DB, DA equal to the given lines; join AB, and draw AC perpendicular to it; DC is the third proportional required. For (V. 8, cor.) BD : DA :: DA : DC.

2. Draw BC perpendicular to AB, and having made BA and AC equal to the less and greater of the given lines, draw CD and BE perpendicular to AC; AD will be a third proportional to AB and AC, and AE to AC and AB. This follows from the third proposition of this book, since the triangles ABC, ACD are equiangular, as are also ABC, AEB.

The angles ABC, ACD, etc., are here made right angles. They may be of any magnitude, however, provided they be equal. It is sufficient, therefore, to draw two straight lines, AB, AC, making any angle; to cut off AB, AC equal to the given proportionals ; and then, BC being joined, to make the angle ACD equal to ABC, and to draw BE parallel to CD; or to make the angle ABE equal to ACB, and to draw CD parallel to BE.

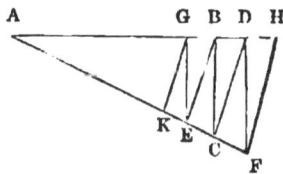

This method affords an easy means of continuing a series of lines in continual proportion, both ways, when any two successive terms are given. Thus, after CD and BE are drawn, it is only necessary to draw DF, EG, etc., parallel to BC, and FH, GK, etc., parallel to CD; as AD, AF, AH, etc., will be the suc-

ceeding terms of the ascending series, and AE, AG, AK, etc.,
those of the descending one.

PROP. X.—PROB.—*To find a fourth proportional to three
given straight lines.*

Let A, B, C be three given straight lines; it is required to
find a fourth proportional to them.

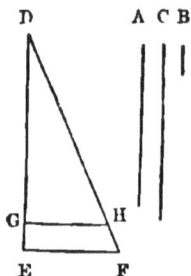

Take two straight lines DE, DF, containing any angle EDF, and make DG equal to
A, GE equal to B, and DH equal to C; and
having joined GH, draw (I. 18) EF parallel
to it through the point E; HF is the fourth
proportional required.

For (V. 2) since GH is parallel to EF, as
DG : GE :: DH : HF; but DG is equal to
A, GE to B, and DH to C; therefore as A :
B :: C : HF; wherefore, to the three given
straight lines, A, B, C, the fourth proportional HF is found;
which was to be done.

Scho. 1. The solution of this problem may also be effected in
several different ways; some of which may be merely indicated
to the student, as the proofs present no difficulty; and it is evident that, with slight modification, they are applicable in solving the ninth proposition, which is only a particular case of the
tenth.

1. Let AE, EC (last fig. to III. 20) be the second and third
terms, placed contiguous, and in the same straight line, and
draw BE, making any angle with AC, and equal to the first
term; through the three points, A, B, C, describe a circle cutting BE produced in D; ED is the fourth proportional. If
AB and CD be joined, the proof will be obtained by means of
the triangles ABE, CDE, which are similar.

2. Draw AB, AC (fig. to III. 21, cor.) making any angle, and
make AB, AE equal to the second and third terms; then if AC
be taken equal to the first term, and a circle be described passing through B, C, and E, and meeting AC in F, AF will be
the required line. If BF and EC be joined, the triangles ABF
and ACE are similar, and hence the proof is immediately obtained.

3. Make BD and DA (fig. to V. 8) perpendicular to each other, and equal to the first and second terms; join AB, and draw AC perpendicular to it; in DA, produced through A, if necessary, take a line equal to the third term, through the upper extremity of which draw a line parallel to AC; the line intercepted on DC, produced if necessary, between D and this parallel, is the fourth proportional required.

Cor. 1. If four straight lines be proportionals, the rectangle contained by the extremes is equivalent to that contained by the means; and (2) if the rectangle contained by the extremes be equivalent to that contained by the means, the four straight lines are proportionals (V. 2, cor.).

Scho. 2. This corollary, of which the corollary immediately following is a case, affords the means of deriving the equality of rectangles, and the proportion of straight lines containing them, from one another. It evidently corresponds to IV. prop. 1, cor., and prop. 2, cor. 1, and it might be regarded as an immediate result of those corollaries, without any distinct proof, if the lines were expressed (I. 23, cor. 4) by lineal units, and the rectangles by superficial ones. This corollary and the third proposition of this book, when employed in connection with one another, form one of the most powerful instruments in geometrical investigations, and they facilitate in a peculiar degree the application of algebra to such inquiries.

Cor. 2. If three straight lines be proportionals, the rectangle contained by the extremes is equivalent to the square of the mean; and (2) if the rectangle contained by the extremes be equivalent to the square of the mean, the three straight lines are proportionals.

PROP. XI.—PROB.—*To find a mean proportional between two given straight lines.*

Let AB, BC be two given straight lines; it is required to find a mean proportional between them.

Place AB, BC in a straight line, and upon AC as diameter describe the semicircle ADC; from B (I. 7) draw BD at right angles to AC; BD is the mean proportional between AB and BC.

Join AD, DC. Then, because the angle ADC in a semicircle

is (III. 11) a right angle, and because in the right-angled triangle ADC, DB is drawn from the right angle perpendicular to AC, DB is a mean proportional (V. 8, cor.) between AB, BC, the segments of the base. Therefore between AB, BC, the mean proportional DB is found; which was to be done.

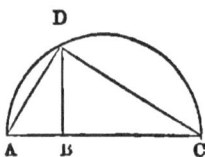

Scho. Out of several additional ways of solving this problem, the following may be mentioned:

1. On the greater extreme as diameter describe a semicircle; from the diameter cut off a segment equal to the less extreme, through the extremity of which draw a perpendicular cutting the circumference; and the chord drawn from that intersection to the extremity of the diameter common to the two extremes is the required mean. The proof is manifest from III. 11, and V. 8, cor.

2. Make AD, DC (2d fig. to III. 21) equal to the given extremes; on AC as chord describe any circle, and a tangent drawn from D will be the required mean. The proof of this is obtained by joining AB and BC, as the triangles ADB, BDC are similar.

When one mean is determined, others may be found between it and the given extremes, and thus three means will be inserted between the given lines; and by finding means between each successive pair of the five terms of which the series then consists, the number of means will be increased to seven. By continuing the process we may find fifteen means, thirty-one means, or any number which is less by one than a power of 2. We cannot find, however, by elementary geometry, any other number of means, such as two, four, or five.

Cor. A given straight line can be divided in extreme and mean ratio.

PROP. XII.—THEOR.—*Equivalent parallelograms which have an angle of the one equal to an angle of the other, have their sides about those angles reciprocally proportional; and* (2) *parallelograms which have an angle of the one equal to an angle of the other, and the sides about those angles reciprocally proportional, are equivalent to one another.*

1. Let AB, BC be equivalent parallelograms, which have the angles at B equal; the sides about those angles are reciprocally proportional; that is, DB : BE :: GB : BF.

Let the sides DB, BE be placed in the same straight line, and conti:nuous, and let the parallelograms be on opposite sides of DE; then (I. 10, cor.) because the angles at B are equal, FB, BG are in one straight line. Complete the parallelogram FE, and (IV. 5) because AB is equal to BC, and that FE is another parallelogram,

AB : FE :: BC : FE.　But (V. 1)

as AB to FE, so is the base DB to BE; and

as BC to FE, so is the base GB to BF;

therefore (IV. 7) as DB : BE :: GB : BF. The sides, therefore, of the parallelograms AB, BC, about their equal angles, are (V. def. 2) reciprocally proportional.

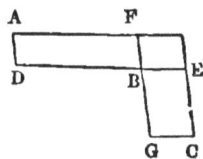

2. But let the sides about the equal angles be reciprocally proportional, viz., DB : BE :: GB : BF; the parallelograms AB, BC are equivalent.

The same construction being made, because,

as DB : BE :: GB : BF; and (V. 1)

as DB : BE :: AB : FE; and as GB : BF :: BC : FE; therefore (IV. 7) as AB : FE :: BC : FE; wherefore (IV. 6) the parallelogram AB is equivalent to the parallelogram BC. Therefore equivalent parallelograms, etc.

Scho. 1. If AD, CG were produced to meet, it would be easy to show, that AB and BC would be the complements of the parallelograms about the diagonal of the whole parallelogram AC.

In the demonstration, it should in strictness be proved that AF and CE meet when produced. This follows from I. 19.

Cor. Hence, equivalent triangles which have an angle of the one equal to an angle of the other, have their sides about those angles reciprocally proportional; and (2) triangles which have an angle of the one equal to an angle of the other, and the sides about those angles reciprocally proportional, are equivalent to one another.

Scho. 2. It is evident from the seventh corollary to the fifteenth proposition of the first book, that this proposition is true as well as when the angles are supplemental as when they are equal.

PROP. XIII.—PROB.—*Upon a given straight line to describe a figure similar to a given rectilineal fi ure, and such that the given line shall be homologous to an assigned side of the given figure.*

Let AB be the given straight line, on which it is required to describe a rectilineal figure similar to a given rectilineal figure, and such that AB may be homologous to CD, a side of the given figure.

First, let the given rectilineal figure be the triangle CDE. Make the angles BAF, ABF respectively equal to DCE, CDE; and (V. 3, cor.) the triangle ABF is similar to CDE, and has AB homologous to CD.

Again: let the given figure be the quadrilateral CDGE. Join DE, and, as in the first part, describe the triangle ABF having the angles BAF, ABF respectively equal to DCE, CDE, and also the triangle BFH having the angles FBH, BFH respectively equal to EDG, DEG. Then (I. 20, cor. 5) the angles BHF, DGE are equal, and (const.) A and C are equal. Also, since (const.) ABF, FBH are respectively equal to CDE, EDG, the whole ABH is equal to the whole CDG. For the same reason AFH is equal to CEG; and therefore the quadrilateral figures ABHF, CDGE are equiangular. But likewise these figures have their sides about the equal angles proportional. For the triangles ABF, CDE being equiangular, and also BFH, DEG; as BA : AF :: DC : CE; and as FH : HB :: EG : GD. Also, in the same triangles, AF : FB :: CE : ED; and as FB : FH :: ED : EG; therefore, *ex æquo,* AF : FH :: CE : EG. In the same manner it may be proved, that AB : BH :: CD : DG; wherefore (V. def. 1) the figures ABHF, CDGE are similar to one another.

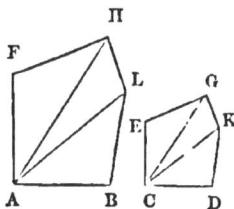

Next: let the given figure be CDKGE. Join DG; and, as in the second case, describe the figure ABHF, similar to CDGE, and similarly situated; also, describe the triangle BHL having the angles BHL, HBL respectively equal to DGK, GDK. Then (1. ax. 2) the whole angles FHL, ABL are respectively equal to the whole angles EGK, CDK; and (I. 20, cor. 5) the angles L and K are equal. Therefore the figures ABLHF, CDKGE are equiangular. Again: because the quadrilaterals ABHF, CDGE, and the triangles BLH, DKG are similar; as FH : HB :: EG : GD, and HB : HL :: GD : GK; therefore, *ex æquo*, FH : HL :: EG : GK. In like manner it may be shown, that AB : BL :: CD : DK; and because the quadrilaterals ABHF, CDGE, and the triangles BLH, DKG are similar, the sides about the angles A and C, L and K, AFH and CEG are proportional. Therefore (V. def. 1) the five-sided figures ABLHF, CDKGE are similar, and the sides AB and CD are homologous. In the same manner, a rectilineal figure of six or more sides may be described on a given straight line, similar to one given; which was to be done.

Scho. In practice, if AB be parallel to CD, the construction is most easily effected by drawing AF and BF parallel to CE and DE; then FH and BH parallel to EG and DG; and lastly, HL and BL parallel to GK and DK. The doing of this is much facilitated by employing the useful instrument, the parallel ruler. For the easiest methods, however, of performing this and many other problems, the student must have recourse to works that treat expressly on such subjects, particularly treatises on practical geometry, surveying, and the use of mathematical instruments.

Prop. XIV.—Theor.—*Similar plane figures are to one another in the duplicate ratio of their homologous sides.*

Let ABC, DEF be similar triangles, having the angles B and E equal, and AB : BC :: DE : EF, so that (IV. def. 13) the side BC is homologous to EF. The triangle ABC has to the triangle DEF the duplicate ratio of that which BC has to EF.

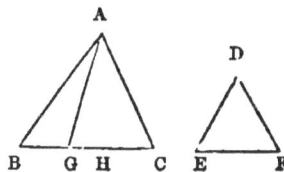

Take (V. 9) BG a third proportional to BC, EF, so that BC : EF :: EF : BG, and join GA. Then, because

AB : BC :: DE : EF ; alternately,

AB : DE :: BC : EF ; but (const.)

as BC : EF :: EF : BG ;

therefore (IV. 7) as AB : DE :: EF : BG. The sides, therefore, of the triangles ABG, DEF which are about the equal angles, are reciprocally proportional, and therefore (V. 12, cor.) the triangles ABG, DEF are equal. Again: because BC : EF :: EF : BG ; and that if three straight lines be proportionals, the first is said (IV. def. 11) to have to the third the duplicate ratio of that which it has to the second ; BC therefore has to BG the duplicate ratio of that which BC has to EF. But (V. 1) as BC to BG, so is the triangle ABC to ABG. Therefore (IV. 7) the triangle ABC has to ABG the duplicate ratio of that which BC has to EF. But the triangle ABG is equal to DEF ; wherefore, also, the triangle ABC has to DEF the duplicate ratio of that which BC has to EF. Therefore, etc.

Scho. 1. From this it is manifest, that if three straight lines be proportionals, as the first is to the third, so is any triangle upon the first to a similar triangle similarly described on the second.

The third proportional might be taken to EF and BC, and placed from E along EF produced ; and then a triangle equal to ABC would be formed by joining D with the extremity of the produced line.

Scho. 2. The above case might also be proved by making BH equal to EF, joining AH, and drawing through H a parallel to AC. The triangle cut off by the parallel is equal (I. 14) to DEF. But (V. 1) the triangle ABC is to the triangle ABH as BC to BH ; and, for the same reason, the triangle ABH is to the triangle cut off by the parallel, or to DEF, as BC to BH, or (const.) as BH to BG. Therefore, *ex æquo*, the triangle ABC has to the triangle DEF the same ratio that BC has to BG, or (IV. def. 11) the duplicate ratio of that which BC has to EF.

Again : let ABCD and AEFG be two squares, then will they be to one another in the duplicate ratio of their sides—as it has been previously demonstrated that similar triangles are to one another in the duplicate ratio of their homologous sides. Therefore :

ABD : AEF in the duplicate ratio of AB to AE ; and ADC :

AFG in the duplicate ratio of AB to AE; hence, by composition, ABD+ADC : AEF+AFG in the duplicate ratio of AB to AE; but ABD +ADC=ABDC, and AEF+AFG= AEFG; wherefore the squares are to one another in the duplicate ratio of their sides.

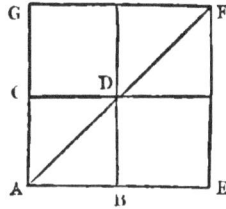

For like reason, the triangles AHD, AFC are to one another in the duplicate ratio that PH is to DF, or that AD is to AC; hence, similar triangles are one to another in the duplicate ratio of their altitudes or bases; since the segments AHD, AFC have the same bases and altitudes as the triangles AHD and AFC, and being segments of quadrants, are similar; hence, they are to one another in the duplicate ratio of their bases and altitudes; therefore, by composition, AED+seg. AHD : ABC+seg. AFC in the duplicate ratio of their homologous sides, or similar polygons are to one another in the duplicate ratio of their homologous sides, and quadrants of circles are one to another in the duplicate ratio of their radii; hence, semicircles are one to another in the duplicate ratio of their diameters, and circles are one to another in the duplicate ratio of their diameters. Wherefore, similar surfaces are, etc.

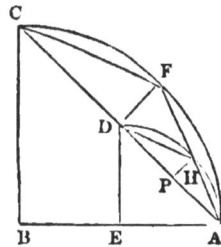

Cor. 1. Therefore, universally, if three lines be proportionals, the first is to the third as any plane figure upon the first to a similar and similarly described figure upon the second.

Cor. 2. Because all squares are similar figures, and all circles are similar figures, the ratio of any two squares to one another is the same as the duplicate ratio of their sides; and the ratio of any two circles to one another is the same as the duplicate ratio of their diameters; hence, any two similar plane figures are to one another as the squares or circles (IV. 7) described on their homologous sides.

Cor. 3. Because the sides of similar plane figures are proportional, therefore (IV. 8) their perimeters or peripheries are proportional to the homologous sides; hence, the perimeters of similar polygons are proportional to their apothems; the circumferences of circles are proportional to their diameters.

Cor. 4. Hence plane figures which are similar to the same figure are similar to one another (I. ax. 1).

Cor. 5. If four straight lines be proportionals, the similar plane figures, similarly described upon them, are also proportionals; and, *conversely*, if the similar plane figures, similarly described upon four straight lines, be proportionals, those lines are proportionals.

Cor. 6. Similar polygons inscribed in circles are to one another as the squares of the diameters.

Cor. 7. When there are three parallelograms, AC, CII, CF, the first, AC (IV. def. 10), has to the third, CF, the ratio which is compounded of the ratio of the first, AC, to the second, CII, and of the ratio of CII to the third, CF; but AC is to CII as their bases; and CII is to CF as their bases; therefore AC has to CF the ratio which is compounded of ratios that are the same with the ratios of the sides.

Scho. 3. Dr. Simson remarks in his Note on this corollary, that "nothing is usually reckoned more difficult in the elements of geometry by learners, than the doctrine of compound ratio." This distinguished geometer, however, has both freed the text of Euclid from the errors introduced by Theon or others, and has explained the subject in such a manner as to remove the difficulties that were formerly felt.   According to him, "every proposition in which compound ratio is made use of, may without it be both enunciated and demonstrated;" and "the use of compound ratio consists wholly in this, that by means of it, circumlocutions may be avoided, and thereby propositions may be more briefly either enunciated or demonstrated, or both may be done.   For instance, if this corollary were to be enunciated, without mentioning compound ratio, it might be done as follows: If two parallelograms be equiangular, and if as a side of the first to a side of the second, so any assumed straight line be made to a second straight line; and as the other side of the first to the other side of the second, so the second straight line be made to a third; the first parallelogram is to the second as the first straight line to the third; and the demonstration would be exactly the same as we now have it.

But the ancient geometers, when they observed this enunciation could be made shorter, by giving a name to the ratio
which the first straight line has to the last, by which name the
intermediate ratios might likewise be signified, of the first to
the second, and of the second to the third, and so on, if there
were more of them, they called this ratio of the first to the last,
the ratio compounded of the ratios of the first to the second,
and of the second to the third straight line; that is, in the
present example, of the ratios which are the same with the
ratios of the sides."

Scho. 4. The seventh corollary will be illustrated by the following proposition, which exhibits the subject in a different,
and, in some respects, a preferable light:

*Triangles which have an angle of the one equal to an angle
of the other, are proportional to the rectangles contained by
the sides about those angles; and (2) equiangular parallelograms are proportional to the rectangles contained by their adjacent sides.*

1. Let ABC, DBE be two triangles, having the angles ABC,
DBE equal; the first triangle is to the second as AB.BC is to
DB.BE.

Let the triangles be placed with their equal angles coinciding,
and join CD.   Then (V. 1) AB is to DB as the triangle ABC
to DBC.   But (V. 1, cor. 2) AB : DB :: AB.BC : DB.BC;
therefore (IV. 7) AB.BC is to DB.BC as the triangle ABC to
DBC.   In the same manner it would
be shown that DB.BC is to DB.BE as
the triangle DBC to DBE; and, therefore, *ex æquo,* AB.BC is to DB.BE as
the triangle ABC to DBE.

2. If parallels to BC through A and
D, and to AB through C and E were
drawn, parallelograms would be formed which would be respectively double of the triangles ABC and DBE, and which
(IV. 9) would have the same ratio as the triangles; that is, the
ratio of AB.BC to DB.BE; and this proves the second part of
the proposition.

Comparing this proposition and the corollary, we see that *the
ratio which is compounded of the ratio of the sides, is the same*

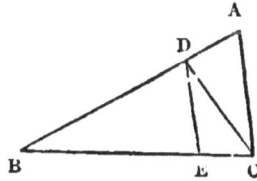

9

as the ratio of their rectangles, or the same (I. 23, cor. 5) *as the ratio of their products, if they be expressed in numbers.* This conclusion might also be derived from the proof given in the text. For (const.) DC : CE :: CG : K; whence (V. 10, cor.) K.DC=CE.CG. But it was proved that BC : K :: AC : CF; or (V. 1) BC.DC : K.DC :: AC : CF; or BC.DC : CE.CG :: AC : CF; because K.DC=CE.CG.

The twelfth proposition of this book is evidently a case of this proposition; and the fourteenth is also easily derived from it.

Prop. XV.—Theor.—*The parallelograms about the diagonal of any parallelogram are similar to the whole, and to one another.*

Let ABCD be a parallelogram, and EG, HK the parallelograms about the diagonal AC; the parallelograms EG, HK are similar to the whole parallelogram, and to one another.

Because DC, GF are parallels, the angles ADC, AGF are (I. 16) equal. For the same reason, because BC, EF are parallels, the angles ABC, AEF are equal; and (I. 15, cor. 1) each of the angles BCD, EFG is equal to the opposite angle DAB, and therefore they are equal to one another; wherefore, in the parallelograms, the angle ABC is equal to AEF, and BAC common to the two triangles BAC, EAF; therefore (V. 3, cor.) as AB : BC :: AE : EF. And (IV. 5), because the opposite sides of parallelograms are equal to one another, AB : AD :: AE : AG; and DC : BC :: GF : EF; and also CD : DA :: FG : GA. Therefore the sides of the parallelograms BD, EG about the equal angles are proportionals; the parallelograms are, therefore (V. def. 1), similar to one another. In the same manner it would be shown that the parallelogram BD is similar to HK. Therefore each of the parallelograms EG, HK is similar to BD. But (V. 14, cor.) rectilineal figures which are similar to the same figure, are similar to one another; therefore the parallelogram EG is similar to HK. Wherefore, etc.

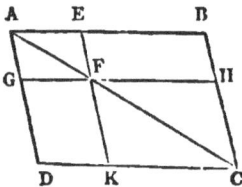

Scho. Hence, GF : FE :: FH : FK. Therefore the sides of

the parallelograms GK and EH, about the equal angles at F, are reciprocally proportional; and (V. 12) these parallelograms are equivalent; a conclusion which agrees with the eighth corollary to the fifteenth proposition of the first book.

PROP. XVI.—PROB.—*To describe a rectilineal figure which shall be similar to one given rectilineal figure, and equivalent to one another.*

Let ABC and D be given rectilineal figures. It is required to describe a figure similar to ABC, and equal to D.

Upon the straight line BC describe (II. 5, scho.) the parallelogram BE equivalent to ABC; also upon CE describe the parallelogram CM equivalent to D, having the angle FCE equal to CBL. Therefore (I. 16 and 10) BC and CF are in a straight line, as also LE and EM. Between BC and CF find (V. 11) a mean proportional GH, and on it describe (V. 13) the figure GHK similar, and similarly situated, to ABC; GHK is the figure required.

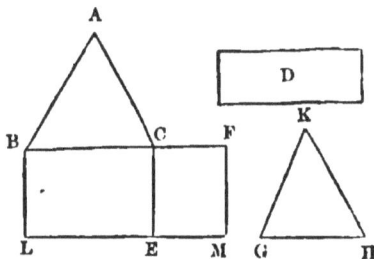

Because BC : GH :: GH : CF, and if three straight lines be proportionals, as the first is to the third, so is (V. 14, cor. 2) the figure upon the first to the similar and similarly described figure upon the second; therefore,

as BC to CF, so is ABC to KGH; but (V. 1)
as BC to CF, so is BE to EF;

therefore (IV. 7) as ABC is to KGH, so is BE to EF. But (const.) ABC is equivalent to BE; therefore KGH is equivalent (IV. ax. 4) to EF; and (const.) EF is equivalent to D; wherefore, also, KGH is equivalent to D; and it is similar to ABC. Therefore the rectilineal figure KGH has been described, similar to ABC and equivalent to D; which was to be done.

PROP. XVII.—THEOR.—*If two similar parallelograms have a common angle, and be similarly situated, they are about the same diagonal.*

Let ABCD, AGEF be two similar parallelograms having a

common angle CAB, they will be about the same diagonal
AD.   Similar parallelograms have
their sides about equal angles proportional (V. def. 1).   Draw the
diagonals EG and CB; hence, AB :
AG :: AC : AE; therefore EG is
parallel to CB (V. 3), and the angles
AEG, ACB are equal (I. 16); likewise the angles EGA, CBA.   The
triangles EAG, FGA are equal (I. 15, cor. 5); likewise the triangles CAB, DBA; therefore AF is equal to EG, and AD is
equal to CB; but AF is the diagonal also of AGEF, and is in
the same straight line with AD, the diagonal of ABCD.
Wherefore, if two similar parallelograms, etc.

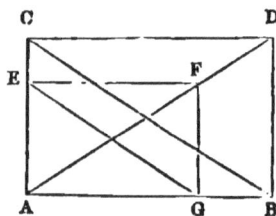

Cor. Hence, equiangular parallelograms have to one another
the ratio which is compounded of the ratio of their sides;
hence, triangles which have one angle of the one equal, or supplemental, to one angle of the other, have to one another the
ratio which is compounded of the ratio of the sides containing
those angles.

PROP. XVIII.—THEOR.— *Of all the parallelograms that can
be inscribed in any triangle, that which is described on the half
of one of the sides as base is the greatest.*

Let ABC be a triangle, having BC, one of its sides, bisected
in D; draw (I. 18) DE parallel to BA, and EF to BC; let also
G be any other point in BC, and
describe the parallelogram GK;
FD is greater than KG.

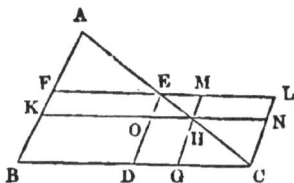

If G be in DC, through C draw
CL parallel to BA, and produce
FE, KH, GH as in the figure.
Then (I. 15, cor. 8) the comple-
ments LH and HD are equivalent; and since the bases CD,
DB are equal, the parallelograms ND, DK (I. 15, cor. 5) are
equivalent.   To LH add ND, and to HD add DK; then (I.
ax. 2) the gnomon MND is equivalent to the parallelogram
KG.   But (I. ax. 9) DL is greater than MND; and therefore

FD, which (I. 15, cor. 5) is equal to DL, is greater than KG, which is equivalent to the gnomon MND. -

If G were in BD, since BD is equal to DC, AE is equal (V. 2) to EC, and AF to FB; and by drawing through A a parallel to BC, meeting DE produced, it would be proved in the same manner that FD is greater than the inscribed parallelogram applied to BG. Therefore, of all the parallelograms, etc.

Cor. Since (V. 15) all parallelograms having one angle coinciding with BCL, and their diagonals with CA, are similar, it follows from this proposition that if, on the segments of a given straight line, BC, two parallelograms of the same altitude be described, one of them, DL, similar to a given parallelogram, the other, DF, will be the greatest possible when the segments of the line are equal.

Scho. The parallelogram FD exceeds KG by the parallelogram OM similar to DL or DF, and described on OH, which is equal to DG, the difference of the bases BD and BG. Hence we can describe parallelograms equivalent and similar to given rectilineal figures.

The enunciation of this proposition here given is much more simple and intelligible than that of Euclid, and the proof is considerably shortened. Euclid's enunciation, as given by Dr. Simson, is as follows: "Of all parallelograms applied to the same straight line, and deficient by parallelograms, similar and similarly situated to that which is described upon the half of the line; that which is applied to the half, and is similar to its defect, is the greatest." It may be remarked, that this proposition, in its simplest case, is the same as the second corollary to the fifth proposition of the second book.

PROP. XIX.—THEOR.—*In equal circles, or in the same circle, angles, whether at the centers or circumferences, have the same ratio as the arcs on which they stand have to one another; so also have the sectors.*

Let, in the equal circles ALB, DNE, the angles LGK, KGC, CGB, DHN, NHM, and MHF be at the centers, and the angles BAC and EDF be at the circumferences, then will those angles have to each other the same ratio as the arcs KL, KC, CB, DN, NM, MF, and FE have to one another.

Since (I. def. 19) all angles at the center of a circle are measured by the arcs intercepted by the sides of the angles,

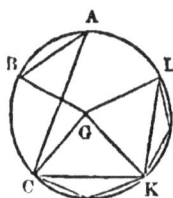

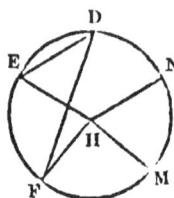

and all angles at the circumference are subtended by the arcs intercepted by the sides of the angles, and (III. 16) equal angles will have equal arcs whether they be at the center or the circumference; hence, the same ratio which the arcs have to each other, will the angles also have to one another—that is, when the arcs be greater, the angles will be greater; less, less; and equal, equal.

And since the sectors are contained (III. def. 7) by the sides of the angles and the arcs, the same ratio between the sectors will evidently exist as there is between the arcs. Wherefore, in equal circles, etc.

Cor. Hence, *conversely*, arcs of the same or equal circles will have the same ratio as the angles or sectors which they measure or subtend—when greater, greater; less, less; or equal, equal.

Prop. XX.—Prob.—*The area of a regular inscribed poly-gon, and that of a regular circumscribed one of the same num-ber of sides being given ; to find the areas of the regular in-scribed and circumscribed polygons having double the number of sides.*

Let A be the center of the circle, BC a side of the inscribed polygon, and DE parallel to BC, a side of the circumscribed one. Draw the perpendicular AFG, and the tangents BH, CK, and join BG ; then BG will be a side

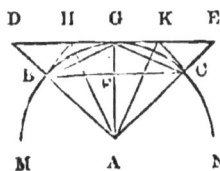

of the inscribed polygon of double the number of sides; and (III. 26, cor. 1) HK is a side of the similar circumscribed one. Now, as a like construction would be made at each of the remaining angles of the polygon, it will be sufficient to con-sider the part here represented, as the tri-angles connected with it are evidently to each other as the poly-gons of which they are parts. For the sake of brevity, then, let P denote the polygon whose side is BC, and P′ that whose side is DE; and, in like manner, let Q and Q′ represent those whose sides are BG and HK; P and P′, therefore, are given; Q and Q′ required.

Now (V. 1) the triangles ABF, ABG are proportional to their bases AF, AG, as they are also to the polygons P′, Q; therefore AF : AG :: P : Q. The triangles ABG, ADG are likewise as their bases AB, AD, or (V. 3) as AF, AG; and they are also as the polygons Q and P′; therefore AF : AG :: Q : P′; wherefore (IV. 7) P : Q :: Q : P′; so that Q is a mean proportional between the given polygons P, P′; and, representing them by numbers, we have $Q^2=PP'$, so that *the area of Q will be computed by multiplying P by P′, and extracting the square root of the product.*

Again: because AH bisects the angle GAD, and because the triangles AHD, AHG are as their bases, we have GH : HD :: AG : AD, or AF : AB :: AHG : AHD. But we have already seen that AF : AG :: P : Q; and therefore (IV. 7) AHG : AHD :: P : Q. Hence (IV. 11) AHG : ADG :: P : P+Q; whence, by doubling the antecedents, 2.AHG : ADG :: 2P : P+Q. But it is evident, that whatever part the triangle ADG is of P′, the same part of the polygon Q′ is the triangle AHK, which is double of the triangle AHG. Hence the last analogy becomes Q′ : P′ :: 2P : P+Q. Now (IV. 2, cor. 1) the product of the extremes is equal to the product of the means; and therefore Q′ will be computed by dividing twice the product of P and P′ by P+Q; and the mode of finding Q has been pointed out already.

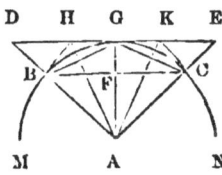

Prop. XXI.—Theor.—*Of regular polygons which have equal perimeters, that which has the greater number of sides is the greater.*

Let AB be half the sides of the polygon which has the less number of sides, and BC a perpendicular to it, which will evidently pass through the center of its inscribed or circumscribed circle; let C be that center, and join AC. Then, ACB will be

the angle at the center subtended by the half side AB. Make
BCD equal to the angle subtended at the center of the other
polygon by half its side, and from C as center, with CD as radius, describe an arc cutting AC in E, and CB produced in F.
Then, it is plain, that the angle ACB is to four right angles as
AB to the common perimeter; and four right angles are to
DCB, as the common perimeter to the half of
a side of the other polygon, which, for brevity, call S; then, *ex æquo*, the angle ACB is
to DCB as AB to S. But (V. 19) the angle
ACB is to DCB as the sector ECF to the
sector DCF; and consequently (IV. 7) the
sector ECF is to DCF as AB to S, and, by
division, the sector ECD is to DCF as AB—
S to S. Now the triangle ACD is greater
than the sector CED, and DCB is less than DCF. But (V. 1)
these triangles are as their bases AD, DB; therefore AD has to
DB a greater ratio than AB—S to S. Hence AB, the sum of
the first and second, has to DB, the second, a greater ratio than
AB, the sum of the third and fourth, has to S, the fourth; and
therefore (IV. 2, cor. 5) S is greater than DB. Let then BG
be equal to S, and draw GH parallel to DC, meeting FC produced in H. Then, since the angles GHB, DCB are equal, BH
is the perpendicular drawn from the center of the polygon having the greater number of sides to one of the sides; and since
this is greater than BC, the like perpendicular in the other
polygon, while the perimeters are equal, it follows that the area
of that which has the greater number of sides is greater than
that of the other.

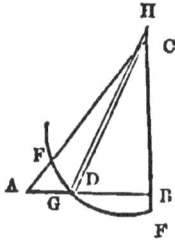

PROP. XXII.—THEOR.—*If the diameter of a circle be divided
into any two parts*, AB, BC, *and if semicircles*, ADB, BEC, *be
described on opposite sides of these, the circle is divided by
their arcs into two figures*, GDE, FED, *the boundary of each of
which is equal to the circumference of* FG ; *and which are such
that* AC : BC :: FG : FED, *and* AC : AB :: FG : GDE.

For (V. 14, cor. 3) the circumferences of circles, and consequently the halves of their circumferences, are to one another as
their diameters; therefore AB is to AC as the arc ADB to

AFC, and BC is to AC as the arc BEC to AFC. Hence (IV.
1) AC is to AC as the compound arc
ADEC to AFC; therefore ADEC is
equal to half the circumference; and the
entire boundaries of the figures GDE,
FED are each equal to the circumference of FG.

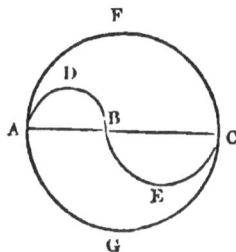

Again (V. 14, cor. 2): circles, and
consequently semicircles, are to one another as the squares of their diameters;
therefore AC^2 is to AB^2 as the semicircle
AFC to the semicircle ADB. Hence, since (II. 4) $AC^2 = AB^2$
$+ 2AB.BC + BC^2$, we find by conversion that AC^2 is to $2AB.$
$BC + BC^2$, as the semicircle AFC to the remaining space
BDAFC; whence, by inversion, $2AB.BC + BC^2$ is to AC^2 as
BDAFC to the semicircle AFC. But BC^2 is to AC^2 as the
semicircle BEC to the semicircle AFC; and therefore (IV. 1)
$2AB.BC + 2BC^2$ is to AC^2 as the compound figure FDE to the
semicircle AFC. But (II. 3) $2AB.BC + 2BC^2 = 2AC.BC$, and
(V. 1) $2AC.BC : AC^2 :: BC : \frac{1}{2}AC$. Hence the preceding
analogy becomes BC to $\frac{1}{2}AC$, as FED to the semicircle, or by
doubling the consequents, and by inversion, AC to BC, as FG
to FED; and it would be proved, in the same manner, that
$AC : AB :: FG : GDE$.

Cor. Hence we can solve the curious problem, in which it is
required to divide a circle into any proposed number of parts,
equal in area and boundary; as it is only necessary to divide
the diameter into the proposed number of equal parts, and to
describe semicircles on opposite sides. Then, whatever part
AB is of AC, the same part is AEG of the circle. Their boundaries are also equal, the boundary of each being equal to the
circumference of the circle.

Scho. Another solution would be obtained, if the circumference were divided into the proposed number of equal parts, and
radii drawn to the points of division. This division, however,
can be effected only in some particular cases by means of elementary geometry.

PROP. XXIII.—PROB.—*To divide a given circle* ABC *into*

any proposed number of equal parts by means of concentric circles.

Divide the radius AD into the proposed number of equal parts, suppose three, in the points E, F, and through these points draw perpendiculars to AD, meeting a semicircle described on it as diameter in G, H; from D as center, at the distances DG, DH, describe the circles GL, HK; their circumferences divide the circle into equal parts.

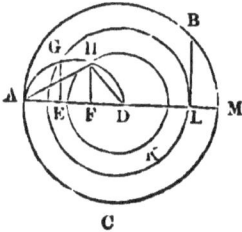

Join AH, DH. Then (V. 17, cor. 2) AD, DH, DF being continual proportionals, AD is to DF as a square described on AD is to one described on DH. But (V. 14, cor. 2) circles are proportional to the squares of their diameters, and consequently to the squares of their radii. Hence (IV. 7) AD is to FD as the circle ABC to the circle HK; and therefore, since FD is a third of AD, HK is a third of ABC. It would be proved in a similar manner that AD is to ED as ABC to GL. But ED is two thirds of AD, and therefore GL is two thirds of ABC; wherefore the space between the circumferences of GL, HK is one third of ABC, as is also the remaining space between the circumferences of ABC and GL.

Cor. Hence it is plain that the area of any *annulus*, or ring, between the circumferences of two concentric circles, such as that between the circumferences of ABC and GL is to the circle ABC as the difference of the squares of the radii DM, DL to the square of DM; or (II. 5, cor. 1, and III. 20) as the rectangle AL.LM, or the square of the perpendicular LB to the square of DM; and it therefore follows (V. 14) that the ring is equivalent to a circle described with a radius equal to LB.

PROP. XXIV.—THEOR.—*If on BC the hypothenuse of a right-angled triangle ABC, a semicircle, BAC, be described on the same side as the triangle, and if semicircles, ADB, AEC, be described on the legs, falling without the triangle, the lunes or crescents ADB, AEC, bounded by the arcs of the semicircles, are together equal to the right-angled triangle ABC.*

For (V. 14 and IV. 9) the semicircle ADB is to the semicircle BAC as the square of AB to the square of BC, and AEC to BAC as the square of AC to the square of BC; whence (IV. 1) the two semicircles ADB, AEC taken together, have to BAC the same ratio as the sum of the squares of AB, AC to the square of BC, that is, the ratio of equality. From these equals take the segments AFB, AGC, and there remain the lunes DF, EG equal to the triangle ABC.

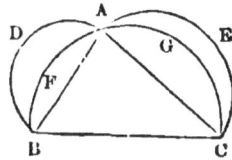

Cor. If the legs AB, AC be equal, the arcs AFB, AGC are equal, and each of them an arc of a quadrant; also the radius drawn from A is perpendicular to BC; and since the halves of equals are equal, each lune is equal to half of the triangle ABC.

·If, therefore, ABC be a quadrant, and on its chord a semicircle be described, the lune comprehended between the circumferences is equal to the triangle ABC; and since (II. 13) a square can be found equal to ABC, we can thus effect the quadrature of a space, ADC, bounded by arcs of circles.

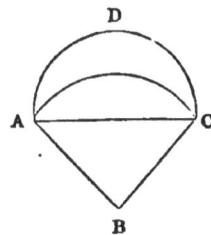

PROP. XXV.—PROB.—*To find the area of a circle.*

Scho. 1. The *approximate* area of a circle can be found by means of the twentieth proposition of this book, by what is called the *method of exhaustions*, giving an error in *excess;* viz., the *approximate* area thus obtained is *square of radius multiplied by* 3.1415926, etc.

Geometry being an exact science, and its conclusions being derived from accurate principles, the *approximate* area for the circle is not consistent with the strictness of geometrical reasoning, and the area of the circle must be established exactly before it can be regarded a geometrical truth. The reason why the *method of e haustions* gives the *approximate* result, is because—by the twentieth proposition of this book—the circumscribed and inscribed regular and similar polygons about the circle are *supposed*, by continually doubling the number of their sides, to be made equivalent to the circle; but CARNOT, in his *Reflexions sur la Metaphysique du Calcul Infinitesimal*, states, " That the ancient geometers did ' not consider it consistent with the strictness of geometrical reasoning to regard curve lines as polygons of a great number of sides." Now, the area of any regular polygon is the rectangle of its apothem and semi-perimeter; but this area is derived from the sixth corollary of the twenty-third proposition of the first book—since it has been shown in the first corollary of the twentieth proposition of the same book, that any rectilineal figure can be divided into as many triangles as the figure has sides; therefore, in case of a regular polygon, when triangles are formed in it by straight lines drawn from the center to the extremities of the several sides of the polygon, the area of the polygon becomes by the tenth axiom of the first book equivalent to the sum of these triangles; hence (I. 23, cor. 6) each triangle is the rectangle of the apothem of the polygon and a semi-side of the polygon; therefore the area of the polygon is (I. ax. 10) the rectangle of its apothem and its semi-perimeter. Since (I. 23, cor. 4) the area of a triangle is derived from the properties of *parallel straight lines*, and any polygon has its sides straight lines (I. def. 12), the properties of *parallel straight lines* are applicable to all polygons; but the circle being formed by a *curve line*, the properties of *parallel straight lines* are not applicable to it; hence the reason is evident why the ancient geometers objected to the curve line being regarded a polygon of a great number of sides. Euclid, in his *Elements*, endeavored to sustain the proposition, that the circle is the rectangle of its radius and semicircumference, by what is called the *indirect, apogogic, or Reductio ad absurdum*, method. Now, every *true* proposition can be *directly* demonstrated, and a fair test of the truth or falsity of this proposition can be in the success or failure of it being *directly* demonstrated. I have given the d rect demonstrations for every other proposition in geometry; but I can not do so in this case—therefore I believe the proposition fallacious. Archimedes has shown that the relation of diameter to the circumference of a circle expressed in numbers, to be as 7 to 22—which is practically correct. Among isoperimetrical figures, the circle contains the greatest area; therefore when 22 expresses the circumference of a circle, the perimeter of its equivalent square must be greater than 22; and if a cube be mechanically constructed upon a base whose perimeter is 24.2487+, it will be equivalent to a cylinder of same height, the diameter of whose base is 7.

Now, when 24.2487+ expresses the perimeter of a square, each of its sides (I. 23, cor. 1) will be 6.0621 + ; and its area will be 36.75, or *three times square of the radius of the circle*. Hence we get by mechanical construction *less* than what is obtained by the *method of exhaustions*. The geometrical confirmation of the mechanical construction is given in the second corollary to the seventeenth proposition of the sixth book.

Scho. 2. Euclid has endeavored to demonstrate that the circle is the rectangle of circumference and semi-radius. Now, the square equal to circle is somewhere between the inscribed and circumscribed squares, and its area is equal to its perimeter multiplied by *less* than semi-radius; consequently the rectangle of circumference and semi-radius will produce more than area of circle.

(*Or Thomson's Euclid, Appendix, Book I., Prop. XXXIX.*)

" The area of a circle is equal to the rectangle under its radius, and a straight line equal to half its circumference. Let AB be the radius of the circle BC; the area of BC is equal to the rectangle under AB and a straight line D equal to half the circumference.

" For if the rectangle AB. D be not equal to the circle BC, it is equal to a circle either greater or less than BC. First, suppose, if possible, the rectangle AB. D to be the area of a circle EF, of which the radius AE is greater than AB." Here Euclid is inconsistent with his own proposition: at the very start he bases his argument upon a contradiction. He premises that the area of a circle is equal to the rectangle under *its* radius, and a straight line equal to half its circumference; then *suppose*, i. e., asks to be granted for the sake of argument, that that same rectangle is equal to a *larger* circle. Why does he resort to this subterfuge? It will be said to show the *Reductio ad absurdum ;* very well, let us follow his argument: "and let GHK be a regular polygon described about the circle BC, such

that its sides do not meet the circumference of EF. Then, by
dividing this polygon into triangles by radii drawn to G, H,
K, etc., it would be seen that its area is equal to the rectangle
under AB and half its perimeter. But the perimeter of the
polygon is greater than the circumference of BC, and therefore
the area of the polygon is greater than the rectangle AB. D;
that is, by hypothesis, than the area of the circle EF, which is
absurd." What is absurd? That the circle EF is greater than
the polygon GHK, etc., or Euclid's argument? The absurdity
is in considering the area of a circle equal to a larger circle.
An argument based upon absurdity must necessarily lead to
absurdity, which in fact has been the case. When Euclid *supposed*, i. e., asked to be granted for the sake of argument, AB.
D=circle EF, it does not prove the area of polygon greater
than the area of circle EF, because he at the start *supposed*
AB. D=circle EF, and consistently with his hypothesis and
his argument, *it must be so to the end ;* therefore, consistently
with his argument and his hypothesis, AB. D is greater than
the area of polygon GHK, etc. The first part of Euclid's proposition is nothing more than a demonstration to prove the area
of a circle is greater than the area of a polygon drawn *within
the circle.* And the second part of Euclid's proposition is
nothing more than a demonstration to prove a circle less than
the circumscribing polygon. This proposition of Euclid is
very sophistical, and consequently its fallacy has been unde-
tected, owing no doubt to the repute of Euclid, and to the supposition that Euclid argued from axioms, and consistently with
the principles of geometry, which he did ; but in this instance
he deceived himself, and consequently all those who believe
him the oracle of geometry. When he attempted to prove
AB. D=area of circle BC, it was contradictory to his argument
to *suppose* AB.'D=area of circle EF ; because when he based
his argument upon the premiss that AB. D=area of circle EF,
consistency demanded that he should stand by his premiss, and
not forsake it as soon as it led to an absurdity, and judge a circle less than a polygon within a circle. The absurdity is in
his own argument, to base it upon a supposition which he knew
was inconsistent with his proposition, and the inconsistence to
drop his premiss when he perceived it led to an absurdity; as

AB. D is less than the circle EF, it is a very fallacious argument, when based on the supposition that they are equal, and it leads to an absurdity; and very inconsistent with geometrical reasoning for Euclid to drop at the conclusion of his argument the very premiss upon which he based his argument. Every method of demonstration, as well as that method termed *Reductio ad absurdum*, require that the premiss which is adopted at the start be retained to the conclusion. And when Euclid adopted AB. D=circle EF at the commencement of his demonstration, consistence of reason and science demanded that he should have kept it to the conclusion, and then there would have been no absurdity, but a demonstration to prove that the polygon GHK, etc., is less than circle EF. But Euclid had in his mind AB. D=circle BC; forgetting that he had adopted AB. D=circle EF, and had *still to prove* AB. D=circle BC.

BOOK SIXTH.

ON THE PLANE AND SOLIDS.

DEFINITIONS.

1. A STRAIGHT line is said to be *perpendicular* to a plane when it makes right angles with all straight lines meeting it in that plane.

2. The *inclination* of two planes which meet one another is the angle contained by two straight lines drawn from any point of their common section at right angles to it, one upon each plane. The angle which one plane makes with another is sometimes called a *dihedral* angle.

3. If that angle be a right angle, the planes are *perpendicular* to one another.

4. *Parallel* planes are such as do not meet one another, though produced ever so far in every direction.

5. A *solid* angle is that which is made by more than two plane angles meeting in one point, and not lying in the same plane.

If the number of plane angles be three, the solid angle is *trihedral ;* if four, *tetrahedral ;* if more than four, *polyhedral.*

6. A *polyhedron* is a solid figure contained by plane figures.

If it be contained by four plane figures, it is called a *tetrahedron ;* if by six, a *hexahedron ;* if by eight, an *octahedron ;* if by twelve, a *dodecahedron ;* if by twenty, an *icosahedron*, etc.

7. A *regular body*, or *regular polyhedron*, is a solid contained by plane figures, which are all equal and similar.

8. Of *solid figures contained by planes*, those are *similar* which have all their solid angles equal, each to each, and which are contained by the same number of similar plane figures, similarly situated.

9. A *pyramid* is a solid figure contained by one plane figure called its *base*, and by three or more triangles meeting in a point without the plane, called the *vertex* of the pyramid.

10. A *prism* is a solid figure, the ends or *bases* of which are parallel, and are equal and similar plane figures, and its other boundaries are parallelograms. One of these parallelograms also is sometimes regarded as the *base* of the prism.

11. Pyramids and prisms are said to be *triangular* when their bases are triangles; *quadrangular*, when their bases are quadrilaterals; *pentagonal*, when they are pentagons, etc.

12. The *altitude of a pyramid* is the perpendicular drawn from its vertex to its base; and the *altitude of a prism* is either a perpendicular drawn from any point in one of its ends or bases to the other; or a perpendicular to one of its bounding parallelograms from a point in the line opposite. The first of these altitudes is sometimes called the *length* of the prism.

13. A *prism*, of which the ends or bases are perpendicular to the other sides, is called a *right prism ;* any other is an *oblique prism.*

14. A *parallelopiped* is a prism of which the bases are parallelograms.

15. A parallelopiped of which the bases and the other sides are rectangles, is said to be *rectangular.*

16. A *cube* is a rectangular parallelopiped, which has all its six sides squares.

17. A *sphere* is a solid figure described by the revolution of a semicircle about its diameter, which remains unmoved.

18. The *axis* of a sphere is the fixed straight line about which the semicircle revolves.

19. The *center* of a sphere is the same as that of the generating semicircle.

20. A *diameter* of a sphere is any straight line which passes through the center, and is terminated both ways by its surface.

21. A *cone* is a solid figure described by the revolution of a right-angled triangle about one of the legs, which remains fixed.

If the fixed leg be equal to the other leg, the cone is called a *right-angled* cone; if it be less than the other leg, an *obtuse-angled,* and if greater, an *acute-angled* cone.

22. The *axis* of a cone is the fixed straight line about which the triangle revolves.

23. The *base* of a cone is the circle described by the leg which revolves.

10

24. A *cylinder* is a solid figure described by the revolution of a rectangle about one of its sides, which remains fixed.

25. The *axis* of a cylinder is the fixed straight line about which the rectangle revolves.

26. The *bases* or *ends* of a cylinder are the circles described by the two revolving opposite sides of the rectangle.

27. *Similar* cones and cylinders are those which have their axes and the diameters of their bases proportionals.

PROPOSITIONS.

PROP. I.—THEOR.— *One part of a straight line can not be in a plane and another part above it.*

Let EFGH be a plane, then the straight line AB will be

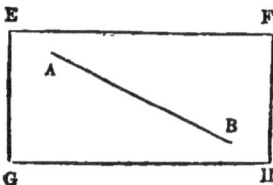

wholly in the plane. By def. 1, Book VI., and def. 7, Book I., AB, being a straight line in the plane EFGH, is wholly in that plane, and can not have one part in the plane and another part above it.

Cor. 1. Hence two straight lines which cut one another are in the same plane; so also are three straight lines which meet one another, not in the same point.

Cor. 2. Hence, if two planes cut one another, their common section is a straight line.

PROP. II.—THEOR.—*If a straight line be perpendicular to each of two straight lines at their point of intersection, it is also perpendicular to the plane in which they are.*

Let the straight line EF be perpendicular to each of the straight lines AB, CD at their intersection E; EF is also perpendicular to the plane passing through AB, CD.

Take the straight lines EB, EC equal to one another, and join BC; in BC and EF take any points G and F, and join EG, FB, FG, FC. Then, in the triangles BEF, CEF, BE is equal to CE; EF common; and the angles BEF, CEF are equal, being (hyp.) right angles; therefore (I. 3) BF is equal to CF. The triangle BFC is therefore isosceles; and (II. 5, cor. 5) the square of BF is equivalent to the square of FG and the rectangle BG.GC. For the same reason, because (const.) the triangle BEC is isosceles, the square of BE is equivalent to the square GE and the rectangle BG.GC. To each of these add the square of EF; then the squares of BE, EF are equivalent to the squares of GE, EF, and the rectangle BG.GC. But (I. 24, cor. 1) the squares of BE, EF are equivalent to the square of BF, because BEF is a right angle; and it has been shown that the square of BF is equivalent to the square of FG and the rectangle BG.GC; therefore the square of FG and the rectangle BG.GC are equivalent to the squares of GE, EF, and the rectangle BG.GC. Take the rectangle BG.GC from each, and there remains the square of FG, equivalent to the squares of GE, EF; wherefore (I. 24, cor.) FEG is a right angle. In the same manner it would be proved that EF is perpendicular to any other straight line drawn through E in the plane passing through AB, CD. But (VI. def. 1) a straight line is perpendicular to a plane when it makes right angles with all straight lines meeting it in that plane; therefore EF is perpendicular to the plane of AB, CD. Wherefore, if a straight line, etc.

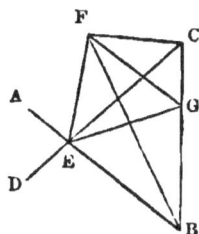

Cor. Hence (VI. def. 1) if three straight lines meet all in one point, and a straight line be perpendicular to each of them at that point, the three straight lines are in the same plane.

PROP. III.—THEOR.—*If two straight lines be perpendicular to the same plane, they are parallel to one another.*

Let the straight lines AB, CD be at right angles to the same plane BDE; AB is parallel to CD.

Let them meet the plane in the points B, D; join BD, and draw DE perpendicular to BD in the plane BDE; make DE equal to AB, and join BE, AE, AD. Then, because AB is perpendicular to the plane, each of the angles ABD, ABE is (VI. def. 1) a right angle. For the same reason, CDB, CDE are right angles. And because AB is equal to

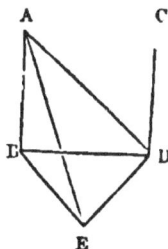

DE, BD common, and the angle ABD equal to BDE, AD is equal (I. 3) to DE.

Again: in the triangles ABE, ADE, AB is equal to DE, BE to AD, and AE common; therefore (I. 4) the angle ABE is equal to EDA; but ABE is a right angle; therefore EDA is also a right angle, and ED perpendicular to DA; it is also perpendicular to each of the two BD, DC; therefore (VI. 2, cor.) these three straight lines DA, DB, DC are all in the same plane. But (VI. 1, cor. 1) AB is in the plane in which are BD, DA; therefore AB, BD, DC are in one plane. Now (hyp.) each of the angles ABD, BDC is a right angle; therefore (I. 16, cor. 1) AB is parallel to CD. Wherefore, etc.

Cor. 1. Hence (I. def. 11) if two straight lines be parallel, the straight line drawn from any point in the one to any point in the other is in the same plane with the parallels.

Cor. 2. Hence, also, if one of two parallel straight lines be perpendicular to a plane, the other is also perpendicular to it.

Also, two straight lines which are each of them parallel to the same straight line, and are not both in the same plane with it, are parallel to one another.

Scho. The same has been proved (I. 17) respecting straight lines in the same plane; therefore, universally, straight lines which are parallel to the same straight line, are parallel to one another.   ·

PROP. IV.—THEOR.—*If two. straight lines meeting one an-other be parallel to two others that meet one another, and are not in the same plane with the first two ; the first two and the other two contain equal angles.*

Let the straight lines AB, BC, which meet one another, be parallel to DE, EF, which also meet one another, but are not in the same plane with AB, BC; the angle ABC is equal to DEF.

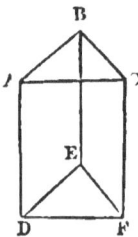

Take BA, BC, ED, EF, all equal to one another, and join AD, CF, BE, AC, DF. Because BA is equal and parallel to ED, therefore AD is (I. 15, cor. 1) both equal and parallel to BE. For the same reason, CF is equal and parallel to BE. Therefore AD and CF being each of them parallel to BE, are (VI. 3, cor. 2) parallel to one another. They are also (I. ax. 1) equal; and AC, DF join them toward the same parts; and therefore (I. 15, cor. 1) AC is equal and parallel to DF. And because AB, BC are equal to DE, EF, and AC to DF, the angle ABC is equal (I. 4) to DEF. Therefore, if two straight lines, etc.

Scho. Or supplemental ones, as will be plain after the demonstration here given, if AB be produced through B. This generalizes the third corollary to the sixteenth proposition of the first book.

Prop. V.—Prob.—*To draw a straight line perpendicular to a plane, from a given point above it.*

Let A be the given point above the plane BII; it is required to draw from A a perpendicular to BII.

In the plane draw any straight line BC, and (I. 8) from A draw AD perpendicular to BC. Then, if AD be also perpendicular to the plane BII, the thing required is done. But if it be not, from D (I. 7) draw DE, in the plane BII, at right angles to BC; from A draw AF perpendicular to DE; and through F draw (I. 18) GII parallel to BC. Then, because BC is at right angles to ED and DA, BC is at right angles (VI. 2) to the plane passing through ED, DA; and GII being parallel to BC, is also (VI. 3, cor. 2) at right angles to the plane through ED, DA; and it is therefore perpendicular (VI. def. 1) to every straight line meeting it in that plane; GII is consequently perpendicular to AF. Therefore AF is perpendicular to each of the straight lines GII, DE; and consequently (VI. 2) to the plane BII; wherefore AF is the perpendicular required.

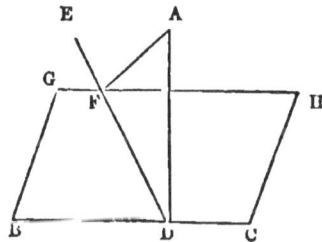

Prop. VI.—Prob.—*To draw a straight line perpendicular to a given plane from a point given in the plane.*

Let A be the point given in the plane; it is required to draw a perpendicular from A to the plane.

From any point B, above the plane, draw (VI. 5) BC perpendicular to it; if this pass through A, it is the perpendicular required. If not, from A draw (I. 18) AD parallel to BC. Then, because AD, CB are parallel, and one of them, BC, is at right angles to the given plane, the other, AD, is also (VI. 3, cor. 2) at right angles to it.

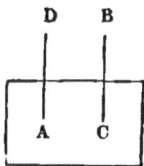

Scho. From the same point in a given plane there can not be two straight lines drawn perpendicular to the plane upon the same side of it; and there can be but one perpendicular to a plane from a point above it.

Cor. Hence planes to which the same straight line is perpendicular, are parallel to one another.

PROP. VII.—THEOR.—*Two planes are parallel, if two straight lines which meet one another on one of them be parallel to two which meet on the other.*

Let the straight lines AB, BC meet on the plane AC, and DE, EF on the plane DF; if AB, BC be parallel to DE, EF, the plane AC is parallel to DF.

From B draw (VI. 3, cor. 2) BG perpendicular to the plane DF, and let it meet that plane in G; and through G draw (I. 18) GH parallel to ED, and GK to EF. Then, because BG is perpendicular to the plane DF, each of the angles BGH, BGK is (VI. def. 1) a right angle; and because (VI. 3, cor. 2) BA is parallel to GH, each of them being parallel to DE, the angles GBA, BGH are together equal (I. 16, cor. 1) to two right angles. But BGH is a right angle; therefore, also, GBA is a right angle, and GB perpendicular to BA. For the same reason, GB is perpendicular to BC. Since, therefore, GB is perpendicular to BA, BC, it is perpendicular (VI. 2) to the plane AC; and (const.) it is perpendicular to the plane DF. But (VI. 6, cor.) planes to which the same straight

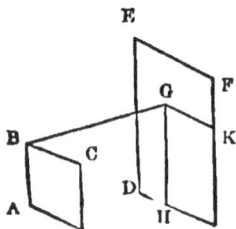

line is perpendicular are parallel to one another; therefore the planes AC, DF are parallel. Wherefore, two planes, etc.

Cor. 1. Hence, if two parallel planes be cut by another plane, their common sections with it are parallels.

Cor. 2. If a straight line be perpendicular to a plane, every plane which passes through it is perpendicular to that plane.

Cor. 3. Hence, if two planes cutting one another be each perpendicular to a third plane, their common section is perpendicular to the same plane.

Prop. VIII.—Theor.—*If two straight lines be cut by parallel planes, they are cut in the same ratio.*

Let the straight lines AB, CD be cut by the parallel planes GH, KL, MN, in the points A, E, B; C, F, D; as AE : EB :: CF : FD.

Join AC, BD, AD, and let AD meet KL in X; join also EX, XF. Because the two parallel planes KL, MN are cut by the plane EBDX, the common sections EX, BD are (VI. 7, cor.) parallel. For the same reason, because GH, KL are cut by the plane AXFC, the common sections AC, XF are parallel. Then (V. 2) because EX is parallel to BD, a side of the triangle ABD, AE : EB :: AX : XD; and because XF is parallel to AC, a side of the triangle ADC, AX : XD :: CF : FD; and it was proved that AX : XD :: AE : EB; therefore (IV. 7) AE : EB :: CF : FD. Wherefore, if two straight lines, etc.

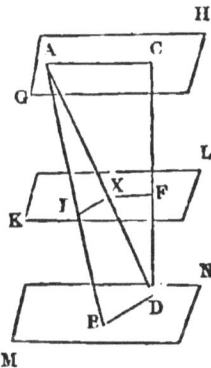

Prop. IX—Theor.—*If a solid angle be contained by three plane angles, any two of them are greater than the third.*

Let the solid angle at A be contained by the three plane angles BAC, CAD, DAB; any two of these are greater than the third.

If the angles be all equal, it is evident that any two of them are greater than the third. But if they be not, let BAC be that angle which is not less than either of the other two, and is greater than one of them, DAB; and make in the plane of BA, AC the angle BAE equal (I. 13) to DAB; make AE equal to AD; through E draw BEC cutting AB, AC in the points B, C, and join DB, DC. Then, in the triangles BAD, BAE, because DA is equal to AE, AB common, and the angle DAB is equal to EAB, DB is equal (I. 3) to BE. Again : because (I. 21, cor.) BD, DC are greater than CB, and one of them, BD, has been proved equal to BE, a part of CB, therefore the other, DC, is greater (I. ax. 5) than the remaining part, EC. Then, because DA is equal (const.) to AE, and AC common, but the base DC greater than the base EC, therefore (I. 21) the angle DAC is greater than EAC, and (const.) the angles DAB, BAE are equal; wherefore (I. ax. 4) the angles DAB, DAC are together greater than BAE, EAC, that is, than BAC. But BAC is not less than either of the angles DAB, DAC; therefore BAC, with either of them, is greater than the other. Wherefore, if a solid angle, etc.

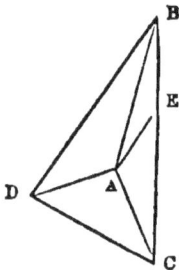

Cor. 1. If every two of three plane angles be greater than the third, and if the straight lines which contain them be all equal, a triangle may be made, having its sides equal, each to each, to the straight lines that join the extremities of those equal straight lines.

Cor. 2. If two solid angles be each contained by three plane angles, equal to one another, each to each ; the planes in which the equal angles are, have the same inclination.

Cor. 3. Two solid angles, contained each by three plane angles which are equal to one another, each to each, and alike situated, are equal to one another.

Cor. 4. Solid figures contained by the same number of equal and similar planes, alike situated, and having none of their solid angles contained by more than three plane angles, are equal and similar to one another.

Cor. 5. If a solid be contained by six planes, two and two of which are parallel, the opposite planes are similar and equal parallelograms.

PROP. X.—THEOR.—*Every solid angle is contained by plane angles, which are together less than four right angles.*

Let the solid angle at A be contained by any number of plane angles, BAC, CAD, DAE, EAF, FAB; these together are less than four right angles.

Let the planes in which the angles are be cut by a plane, and let the common sections of it with those planes be BC, CD, DE, EF, FB. Then, because the solid angle at B is contained by three plane angles, CBA, ABF, FBC, of which (VI. 9) any two are greater than the third, CBA, ABF are greater than FBC. For the same reason, the two plane angles at each of the points C, D, E, F, viz., the angles which are at the bases of the triangles having the common vertex A, are greater than the third angle at the same point, which is one of the angles of the figure BCDEF. Therefore all the angles at the bases of the triangles are together greater than all the angles of that figure; and because (I. 20) all the angles of the triangles are together equal to twice as many right angles as there are triangles—that is, as there are sides in the figure BCDEF; and that (I. 20, cor. 1) all the angles of the figure, together with four right angles, are likewise equal to twice as many right angles as there are sides in the figure; therefore all the angles of the triangle are equal to all the angles of the figure, together with four right angles. But all the angles at the bases of the triangles are greater than all the angles of the figure, as has been proved; wherefore the remaining angles of the triangles, viz., those at the vertex, which contain the solid angle at A, are less than four right angles. Therefore, every solid angle, etc.

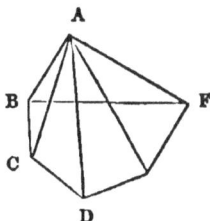

Scho. This proposition does not necessarily hold, if any of the angles of the rectilineal figure BCDEF be re-entrant; or, which is the same, if any of the planes forming the solid angle at A, being produced, pass through that angle.

PROP. XI.—PROB.—*To make a solid angle having the angles containing it equal to three given plane angles, any two of*

*which are greater than the third, and all three together less than
four right angles.*

Let B, E, H be given plane angles, any two of which are
greater than the third, and all of them together less than four
right angles; it is required to make a solid angle contained by
plane angles equal to B, E, H, each to each.

From the lines containing the angles, cut off BA, BC, ED,
EF, HG, HK, all equal to one another, and join AC, DF, GK;
then (VI. 9, cor. 1) a triangle may be made of three straight
lines equal to AC, DF, GK. Let this (I. 12) be the triangle
LMN, AC being equal to LM, DF to MN, and GK to LN.

About LMN describe (III. 25, cor. 2) a circle, and draw the
radii, LO, MO, NO; draw also OP (VI. 6, cor.) perpendicular to the plane LMN. Then, any of the radii LO, MO, NO is
less than AB. Find (I. 24, cor. 3) the side of a square equivalent to the difference of the squares of AB and LO; make OP
equal to that side, and join PL, PM, PN; the plane angles
LPM, MPN, and NPL form the solid angle required.

For, since OP is (const.) perpendicular to the plane LMN,
the angles LOP, MOP, NOP are (VI. def. 1) right angles;
and therefore, since in the triangles OLP, OMP, ONP the
sides OL, OM, ON are equal, OP common, and the contained

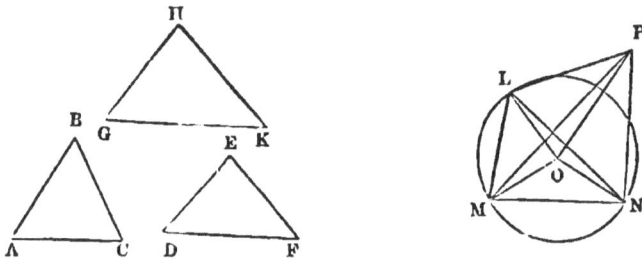

angles equal, the bases LP, MP, NP are (I. 3) all equal. Also
(const.) the square of AB is equivalent to the squares of LO,
OP; and (I. 24, cor. 1) the square of LP is also equivalent to
the squares of LO, OP, because LOP is a right angle. Therefore (I. ax. 1) the square of AB is equal to the square of PL,
and (I. 23, cor. 3) AB to PL; and hence all the straight lines
LP, MP, NP, BA, BC, ED, etc., are equal. Then, in the triangles LPM, ABC, the sides AB, BC are equal to LP, PM,

each to each ; and (const.) AC is equal to LM ; therefore (I. 4) the angles ABC, LPM are equal; and it would be shown in a similar manner, that the angle E is equal to MPN, and H to NPL. The solid angle at P, therefore, being contained by three plane angles, which are equal to the three given angles, B, E, H, each to each, is such as was required.

PROP. XII.—THEOR.—*A plane cutting a solid, and parallel to two of its opposite planes, divides the whole into two solids, the base of one of which is to the base of the other as the one solid is to the other.*

Let the solid BC be cut by the plane GF which is parallel to the opposite planes BY and HC, and divides the whole into two solids, BF and GC ; as the base of the first is to the base of the second, so is BF to GC.

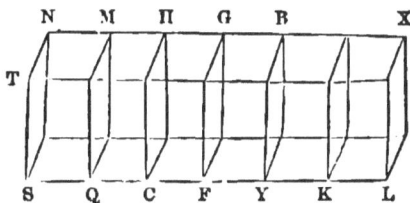

Produce YC both ways, and take on one side any number of straight lines, YK and KL, each equal to YF, and complete parallelograms similar and equal to BY. Then, because BY, YK, and KL are all equal, the parallelograms on them are also equal (I. 15, cor. 5) ; and for same reason the parallelograms on the other side of YC, on the straight lines CQ, QS, each equal to FC, are also equal ; therefore three planes of the solid XK are equal and similar to three planes of KB, as also to three planes of YG. But (VI. 9, cor. 5) the three planes opposite to these three are equal and similar to them in the several solids ; and none of their solid angles are contained by more than three plane angles ; therefore (VI. 9, cor. 4) the solids XK, KB, and YC are equal. For the same reason, FH, CM and QN are also equal. Therefore whatever multiple the base LF is of YF, the same multiple is the solid LG of YG. For same reason, whatever multiple the base FS is of FC, the same

multiple is the solid FT of FII. And if the base LF be equal
to SF, the solid LG is equal (VI. 9, cor. 4) to FT; if greater,
greater; and if less, less. Therefore (IV. def. 5) as the base
YF is to the base FC, so is the solid BF to the solid GC;
wherefore, a plane, etc.

PROP. XIII.—PROB.—*At a given point in a given straight
line, to make a solid angle equal to a given solid angle contained by three plane angles.*

Let A be a given point in a given straight line AB, and D a
given solid angle contained by the three plane angles EDC,
EDF, FDC; it is required to make at A in the straight line
AB a solid angle equal to the solid angle D.

In DF take any point F, from which draw (VI. 5) FG perpendicular to the plane EDC, meeting that plane in G; join
DG, and (I. 13) make the angle BAL equal to EDC, and in
the plane BAL make the angle BAK equal to EDG; then
make AK equal to DG, and (VI. 6) draw KII perpendicular to
the plane BAL, and equal to GF, and join AII. Then the
solid angle at A, which is contained by the plane angles BAL,
BAII, IIAL, is equal to the given solid angle at D.

Take AB DE equal to one another; and join IIB, KB, FE,
GE; and (VI. def. 1) because FG is perpendicular to the plane
EDC, FGD, FGE are right angles. For the same reason,
IIKA, IIKB are right angles; and because KA, AB are equal
to GD, DE, each to each, and contain equal angles, BK is
equal (I. 3) to EG; also KII is equal to GF, and IIKB, FGE
are right angles; therefore IIB is equal to FE. Again: because AK, KII are equal to DG, GF, and contain right angles,

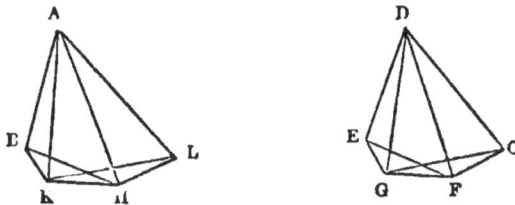

AII is equal to DF; also AB is equal to DE, and IIB to FE;
therefore (I. 4) the angles BAII, EDF are equal. Again: since

(const.) the angle BAL is equal to EDC, and BAK to EDG, the remaining angles KAL, GDC are (I. ax. 3) equal to one another; and, by taking AL and DC equal, and joining LH, LK, CF, CG, it would be proved, as in the foregoing part, that the angle HAL is equal (I. 4) to FDC. Therefore, because the three plane angles BAL, BAH, HAL, which contain the solid angle at A, are equal to the three EDC, EDF, FDC, which contain the solid angle at D, each to each, and are situated in the same order, the solid angle at A is equal (VI. 9, cor. 3) to the solid angle at D. Therefore, what was required has been done.

PROP. XIV.—THEOR.—*If a parallelopiped be cut by a plane passing through the diagonals of two of the opposite planes, it is bisected by that plane.*

Let AB be a parallelopiped, and DE, CF the diagonals of the opposite parallelograms AH, GB, viz., those which join the equal angles in each. Then (VI. 3, cor. 2) CD, FE are parallels, because each of them is parallel to GA; wherefore (VI. 3, cor. 1) the diagonals CF, DE are in the plane in which the parallels are, and (VI. 8) are themselves parallels. Again: because (I. 15, cor. 1) the triangle CGF is equal to CBF, and DAE to DHE; and that (VI. 12) the parallelogram CA is equal and similar to the opposite one BE; and GE to CH; therefore the prism contained by the two triangles CGF, DAE, and the three parallelograms CA, GE, EC, is equal (VI. 9, cor. 4) to the prism contained by the two triangles CBF, DHE, and the three parallelograms BE, CH, EC; because they are contained by the same number of equal and similar planes, alike situated, and none of their solid angles are contained by more than three plane angles. Therefore, if a parallelopiped, etc.

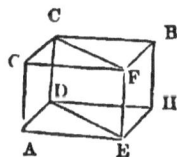

Scho. The *insisting lines* of a parallelopiped are the sides of the parallelograms between the base and the opposite plane.

Cor. In a parallelopiped, if the sides of two of the opposite planes be each bisected, the common section of the planes passing through the points of division, and any diagonal of the parallelopiped bisect each other.

PROP. XV.—THEOR.—*Parallelopipeds upon the same base, and of the same altitude, the insisting lines of which are terminated in the same straight lines of the plane opposite to the base, are equal to one another.*

Let the parallelopipeds AH, AK (2d fig.) be upon the same base AB, and of the same altitude; and let their insisting lines AF, AG, LM, LN be terminated in the same straight line FN, and CD, CE, BH, BK in the same DK; AH is equal to AK.

First, let the parallelograms DG, HN, which are opposite to the base AB, have a common side HG. Then because AH is cut by the plane AGHC passing through the diagonals AG, CH of the opposite planes ALGF, CBHD, AH is bisected (VI. 14) by the plane AGHC. For the same reason, AK is bisected by the plane LGHB through the diagonals LG, BH. Therefore the solids AH, AK are equal, each of them being double of the prism contained between the triangles ALG, CBH.

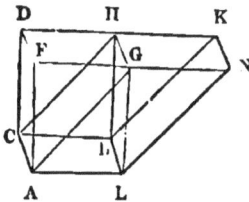

But let the parallelograms DM, EN, opposite to the base, have no common side. Then, because CH, CK are parallelograms, CB is equal (I. 15, cor. 1) to each of the opposite sides DH, EK; wherefore DH is equal to EK. From DK take separately EK, DH; then DE is equal to HK; wherefore, also (I. 15, cor. 5), the triangles CDE, BHK are equal; and (I. 15, cor. 5) the parallelogram DG is equal to HN. For the same reason, the triangle AFG is equal to LMN; and (VI. 12) the parallelogram CF is equal to BM, and CG to BN; for they are opposite. Therefore (VI. 9, cor. 4) the prism which is contained by the two triangles AFG, CDE, and the three parallelograms AD, DG, GC, is equal to the prism contained by the two triangles LMN, BHK, and the three parallelograms BM, MK, KL. If, therefore, the prism LMNBHK be taken from the solid of which the base is the parallelogram AB, and in which FDKN is the one opposite to it; and if from the same solid there be

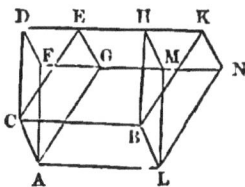

taken the prism AFGCDE, the remaining solids AII, AK are equal. Therefore, parallelopipeds, etc.

Cor. 1. Also parallelopipeds upon the same base and of the same altitude, the insisting lines of which are not terminated in the same straight lines in the plane opposite to the base, are equal to one another.

Cor. 2. Hence parallelopipeds which are upon equal bases, and of the same altitude, are equal to one another (I. 15, cor. 5).

Cor. 3. Parallelopipeds which have the same altitude, are to one another as their bases (VI. 12).

Cor. 4. If there be two triangular prisms of the same altitude, the base of one of which is a parallelogram, and that of the other a triangle; if the parallelogram be double of the triangle, the prisms are equal (I. 15, cor. 4).

Prop. XVI.—Theor.—*Similar solids are one to another in the triplicate ratio of their homologous sides.*

Let AB, CD be similar parallelopipeds, and the side AE homologous to CF; AB has to CD the triplicate ratio of that which AE has to CF.

Produce AE, GE, IIE, and in these produced take EK equal to CF, EL equal to FN, and EM equal to FR; and complete the parallelogram KL and the solid KO. Because KE, EL are equal to CF, FN, and the angle KEL equal to the angle CFN, since it is equal to the angle AEG, which is equal to CFN, because the solids AB, CD are similar; therefore the parallelogram KL is similar and equal to CN. For the same reason, the parallelogram MK is similar and equal to CR, and also OE to FD. Therefore three parallelograms of the solid KO are equal and similar to three parallelograms of the solid CD; and (VI. 12) the three opposite ones in each solid are equal and similar to these. Therefore (VI. 9, cor. 4) the solid KO is equal and similar to CD. Com-

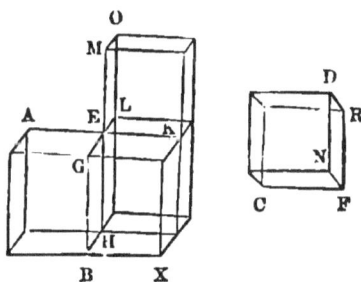

plete the parallelogram GK, and the solids EX, LP upon the
bases GK, KL. so that EH may be an insisting line in each of
them; and thus they are of the same altitude with the solid
AB. Then, because the sol-
ids AB, CD are similar (VI.
def. 8, and alternately), as
AE is to CF, so is EG to FN,
and so is EH to FR; and FC
is equal to EK, and FN to
EL, and FR to EM; there-
fore, as AE to EK, so is EG
to EL; and so is HE to EM.
But (V. 1) as AE to EK, so
is the parallelogram AG to GK; and as GE to EL, so is GK
to KL; and as HE to EM, so is BE to KM; therefore (IV. 7)
as the parallelogram AG to GK, so is GK to KL, and PE to
KM. But (VI. 13) as AG to GK, so is the solid AB to EX;
and as GK to KL, so is the solid EX to PL; and as PE to
KM, so is the solid PL to KO; and therefore (IV. 7) as the
solid AB to EX, so is EX to PL, and PL to KO. But if four
magnitudes be continual proportionals, the first is said to have
to the fourth the triplicate ratio of that which it has to the second; therefore the solid AB has to KO the triplicate ratio of
that which AB has to EX. But as AB is to EX, so is the parallelogram AG to GK, and the straight line AE to EK.
Wherefore the solid AB has to the solid KO the triplicate ratio
of that which AE has to EK; and the solid KO is equal to CD,
and the straight line EK to CF. Therefore the solid AB has
to CD the triplicate ratio of that which the side AE has to the
homologous side CF.

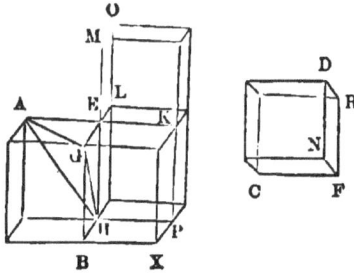

Since (VI. 15) CNF is half of the base of CD, CRF half of CR,
and NRF half of NR, which planes form the triangular pyramid
CNF, R; and since the triangular pyramid AGE, H is formed in
a similar manner, and the parallelopiped AB has to the parallelopiped CD the triplicate ratio of that which the side AE
has to the side CF; hence (IV. ax. 1) the triangular pyramid
CNF, R has to the triangular pyramid AGE, H the triplicate
ratio of that which AE has to CF. And (I. 20, cor. 1) all
polygons can be divided into triangles; therefore (V. 14) solids

on similar bases have to one another the triplicate ratio of homologous sides. Wherefore, similar solids are, etc.

Cor. 1. Hence, similar solids of the same or equal bases are to one another as their altitudes—and, conversely, those of the same or equal altitudes are to one another as their bases.

Cor. 2. Hence, also, the bases and altitudes of equivalent solids are reciprocally proportional; and conversely, solids having their bases and altitudes reciprocally proportional, are equivalent.

Cor. 3. Hence, cones and cylinders upon equal bases are as their altitudes, and their bases and altitudes are reciprocally proportional when the cones and cylinders are equivalent. And similar cones and cylinders have to one another the triplicate ratio of that which the diameters of their bases have; and spheres have the triplicate ratio of their diameters (V. 14 and VI. 16).

Cor. 4. From this it is manifest, that if four straight lines be continual proportionals, as the first is to the fourth, so is the parallelopiped described from the first to the similar solid similarly described from the second; because the first straight line has to the fourth the triplicate ratio of that which it has to the second.

Cor. 5. Parallelopipeds contained by parallelograms equiangular to one another, each to each, that is, of which the solid angles are equal, each to each, have to one another the ratio which is the same with the ratio compounded of the ratios of their sides (V. 14, cor. 7).

Cor. 6. The bases and altitudes of equivalent parallelopipeds are reciprocally proportional; and (2) if the bases and altitudes be reciprocally proportional, the parallelopipeds are equivalent.

PROP. XVII.—THEOR.—*Every pyramid is one third the prism of the same base and altitude, and every cone is one third of the cylinder with the same base and altitude, or every pyramidal solid is one third the solid of the same base and altitude.*

Let there be a prism of which the bases are the triangles ABC, DEF; the prism may be divided into three equal triangular pyramids.

11

Join BD, EC, CD; and because ABED is a parallelogram, and BD its diagonal, the triangles ABD, EBD are (I. 15, cor. 1) equal; therefore (VI. 16, cor. 3) the pyramid of which the base is ABD, and vertex C, is equal to the pyramid of which the base is EBD and vertex C. But EBC may be taken as the base of this pyramid, and D as its vertex. It is therefore equal (VI. 16, cor. 3) to the pyramid of which ECF is the base, and D the vertex; for they have the same altitude, and (I. 15, cor. 1) equal bases ECF, ECB; and it has been already proved to be equal to the pyramid ABDC. Therefore the prism ABCDEF is divided into three equal pyramids having triangular bases, viz., into the pyramids ABDC, EBDC, ECFD. Therefore, every triangular prism, etc.

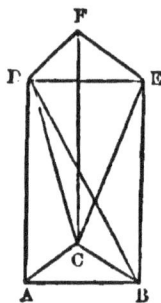

Now every polygon can be divided into triangles (I. 20, cor. 1); hence every prism with a polygonal base can be divided into prisms having triangular bases; and as each of these triangular prisms can be divided into three equivalent triangular pyramids, therefore every pyramid is one third the prism of same base and altitude. As it has been shown (V. 14) that all surfaces are to one another in the duplicate ratio of their homologous sides, and (VI. 16) all solids are to one another in the triplicate ratio of their homologous sides, it follows that all solids of similar bases and altitudes have the same proportion to one another (VI. 16, cor. 3); hence, cones having similar bases and altitudes to cylinders, have the same proportion to those cylinders which pyramids have to prisms of similar bases and altitudes; therefore the cones are one third the cylinders, as it is evident that the section of the cone and cylinder is similar to the section of the pyramid and prism of whatever regular and similar base. For ABC can be the section of a cone and of a pyramid of any regular base, and ABED can be the section of a prism of any similar base, and section of a cylinder of similar base with cone; therefore the same proportion which regulates the respective magnitudes

of the pyramid and prism, also regulates the respective magnitudes of cone and cylinder—as all surfaces are (V. 14) to one another in the duplicate ratio of their homologous sides, and (VI. 16) all solids are to one another in the triplicate ratio of their homologous sides.

Cor. 1. Every sphere is two thirds of its circumscribing cylinder. Let ABCD be a cylinder circumscribing the sphere EFGH; then will the sphere EFGH be two thirds of the cylinder ABCD. For, let the plane AC be a section of the sphere and cylinder through the center J. Join AJ, BJ. Also, let FJH be parallel to AD or BC, and OKL be parallel to AB or DC, the base of the cylinder, the latter line, KL, meeting BJ in M, and the circular section of the sphere in N.

Then, if the whole plane HFBC be conceived to revolve about the line HF as an axis, the square FG will describe a cylinder, AG; the quadrant JFG will describe a hemisphere, EFG; and the triangle JFB will describe a cone, JAB. Also, in the rotation, the three lines or parts KL, KN, KM, as radii, will describe corresponding circular sections of those solids, viz., KL, a section of the cylinder; KN, a section of the sphere; and KM, a section of the cone. Now, FB being equal to FJ or JG, and KL parallel to FB, then by similar triangles JK is equal to KM. And since in the right-angled triangle JKN, $JN^2 = JK^2 + KN^2$ (I. 24, cor. 1), and because KL is equal to the radius JG or JN, and KM is equal to JK, therefore $KL^2 = KM^2 + KN^2$; and because circles are as the squares of their diameters or the squares of their radii, therefore (V. 14, cor. 2) circle of KL is equivalent to circles of KM and KN, or section of cylinder is equivalent to both corresponding sections of sphere and cone. And as this will always be, it follows that the cylinder EB, which is all the former sections, is equivalent to the hemisphere EFG and cone JAB, which are all the latter sections. But JAB is one third of the cylinder EB (VI. 17); therefore (I. ax. 3) the hemisphere EFG is two thirds of the cylinder EB.

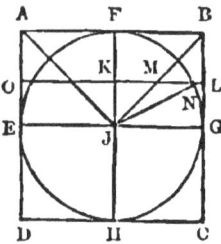

Cor. 2. If the parallelogram BEGC be revolved around the

fixed axis BC, it will generate a cylinder (VI. def. 24); the semicircle BNC will generate a sphere (VI. def. 17); and the triangle BGC will generate a cone (VI. def. 21). The cone will be one third the cylinder (VI. 17), and the sphere will be two thirds the same cylinder (VI. 17, cor 1).

The triangle BOP having one half altitude and one half base of the triangle BGC, will generate a cone one eighth of the cone generated by the triangle BGC (VI. 16, cor. 3); hence, one twelfth of the cylinder generated by the square BENP; and the cone generated by the triangle BNP is one half cone generated by the triangle BGC (VI. 16, cor. 1); hence, four times cone generated by the triangle BOP. And the hemisphere generated by the quadrant BNP is two thirds cylinder generated by the square BENP VI. 17, cor. 1), or eight times cone generated by the triangle BOP.

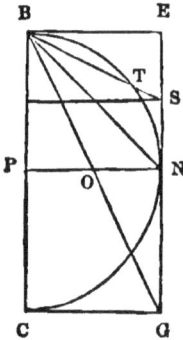

Let the triangle BSN be described on BN, equal to the triangle BON (I. 23, and 15, cor. 4). Then the trapezium BSNP will generate a solid equivalent to the sum of a cylinder one half cylinder generated by the square BENP, and a cone one sixth of the same cylinder, or eight times the cone generated by the triangle BOP, making a solid equivalent to the hemisphere generated by the quadrant BNP on the same radius PN and same altitude BP. But the triangle BNP is common to both the trapezium BSNP and the quadrant BNP, and generates in each case the solid equivalent to four times cone generated by the triangle BOP; therefore the segment BN and the triangle BSN generate an equivalence of solid, or four times cone generated by the triangle BOP; consequently the segment BN and the triangle BSN are equivalent (I. ax. 1).

Again: the triangle BNP generates a cone one third the cylinder generated by the square BENP (VI. 17), and the quadrant BNP generates a hemisphere two thirds of the same cylinder (VI. 17, cor. 1). The triangle BNP is one half the square BENP.

Now, the trapezium BSNP, equivalent to three fourths of the

square, on same radius and altitude as the square, generates
a solid two thirds of the solid generated by the square, and the
quadrant BNP with same radius and altitude as the square
BENP generates an equivalent solid with the trapezium
BSNP. That an equivalence of surfaces upon the same radius
will generate an equivalence of solids can be illustrated by
taking a trapezium greater than the trapezium BSNP, having
the same radius. It can easily be shown that the greater trapezium generates a greater solid than the less trapezium, and in
a similar manner it can be shown that a less trapezium than the
trapezium BSNP generates a less solid; hence, very evidently,
when a greater surface upon same radius generates a greater
solid, and a less surface generates a less solid, equivalent
surfaces must generate equivalent solids on the same radius;
and, *conversely*, when we have equivalent solids generated
upon the same radius, the generating surfaces are equivalent;
therefore (I. ax. 1) the quadrant BNP is three fourths of the
square BENP, or the semicircle BNC is three fourths of the
parallelogram BEGC, or any circle is three fourths of the circumscribing square, or THREE TIMES SQUARE OF RADIUS.
Hence, we have a geometrical confirmation of the *mechanical
construction* in scholium to twenty-fifth proposition of book
fifth.

OTHERWISE :

The triangle BGC generates a cone *one third* (VI. 17) of
the cylinder generated by the rectangle BEGC, and the semicircle BNC generates a sphere *two thirds* (VI. 17, cor. 1) of the
same cylinder; the sphere is the *mean* between the cone and
cylinder; therefore the semicircle is evidently the *mean* between the triangle BGC and the rectangle BEGC, or *three
fourths* of the rectangle BEGC; or any circle is three fourths
square of its diameter, or *three times square of its radius*.

OTHERWISE :

Circles are to one another as the squares described on their
diameters (V. 14); consequently squares are to one another as
the circles described on their sides; therefore there is an
equality of proportion (V. 24); hence, we derived the arithmetical proportion :

Rectangle BEGC, semicircle BNC, triangle BGC.

The sum of extremes is equivalent to twice the mean; therefore we have—

Rectangle BEGC+triangle BGC$=\bigcirc 2$ semicircle BNC; or, semicircle BNC$=\bigcirc \frac{1}{2}$ rectangle BEGC$+\frac{1}{2}$ triangle BGC.

Also, the difference between first and second terms of an arithmetical proportion is the same as the difference between the second and third terms, as the difference between third and fourth terms, and so on; hence we have—

Rectangle BEGC−semicircle BNC$=\bigcirc$semicircle BNC−triangle BGC; therefore we get segment BN$=\bigcirc$triangle BSN, or one fourth square BENP; consequently, circle $=\bigcirc$ three fourth square of diameter, or three times square of radius.

Cor. 3. Archimedes discovered the proportion 1, 2, 3 between the cone, sphere, and cylinder of similar dimensions; but from the previous corollary we obtain the proportion 1, 2, 3, 4 for the cone, sphere, cylinder, and *cube* of similar dimensions; because the cube is *eight times cube of radius of the sphere* (VI. 16); the cylinder is *six times cube of radius of the sphere* (VI. 17, cor. 2); the sphere is *four times cube of radius of the sphere* (VI. 17, cors. 1 and 2); and the cone is *twice cube of radius of the sphere* (VI. 17, cors. 1 and 2). Hence the sphere is the *mean proportional* of the cone, and the cube circumscribing the sphere, or *one half* the circumscribing cube; therefore the *surface* of the sphere is *four times the area of one of its great circles*, or *two thirds* the surface of the circumscribing cylinder. Hence, there is the identical proportion between the *surfaces* of the sphere and cylinder as there is between the *solidities* of the sphere and cylinder.

Scho. 1. Therefore the second corollary gives the solution to the long mooted and much vexed question of the Quadrature of the Circle, showing that the perplexity of it arose from the *ungeometrical supposition* (V. 25, scho.) that "the circle is a regular polygon of an infinite number of sides." Hence it is evident that all conclusions derived from a fallacious supposition will give perplexity so long as the supposition is maintained, and must necessarily involve contradictions to the rigor of geometrical reasoning. And when demonstrations are conducted consistently with established definitions, axioms, and propositions, all conclusions derived from them are unimpeachable, and are valuable to a system of scientific truths.

Scho. 2. Geometry, like all other sciences, is based upon certain *fundamental* principles, and a close examination of this science reveals the fact that, throughout its whole extent and its various applications, the principle that the *straight line is the shortest line between two given points* is the *fundamental* principle of the science; by this principle the dimensions of magnitudes are determined, distances of objects made known, and other useful and practical results ascertained. Since this principle is so important, it would be interesting to inquire the reason why it has such manifest usefulness. The angle is a magnitude contained by the intersection of two *straight lines*, and the polygon is another magnitude bounded by three or more *straight lines;* hence we see how intimate the connection between the *straight line* and the angle and polygon; therefore we find that the *functions* of the angle are *straight lines*, such as sines, co-sines, tangents, etc.; and the properties of the polygon are defined by *straight lines*, such as its perimeter and apothem; therefore in all rectilineal magnitudes we discover a use for the *straight line* above all other lines, and evidently the principle of the straight line has a peculiar force to all rectilineal figures; consequently we adopt the *straight line* as a *means of measure* for all rectilineal magnitudes. The adoption of this *means of measure* constitutes the *straight line* a *standard* by which all measurements of rectilineal magnitudes are compared. Hence very naturally there is a consistency between the measurements and other properties of rectilineal magnitudes.

Now, when we examine the circle or portions of the circle, as the segments, sectors, arcs, etc., we at once discover a *non-coincidence* between the *curve* which bounds them and the *straight line* which bounds rectilineal magnitudes; hence, very evidently, the *superficies* of curvilinear spaces require a peculiar connection between them and the bounding *curve*, as there is a peculiar connection between the *superficies* of rectilineal spaces and the bounding *straight lines*. In other words, since we derive the measurements and other properties of rectilineal magnitudes from the principle of the *straight line*, so we must determine the measurements and other properties of curvilinear magnitudes from the principle of the *curve*, and thus we perceive *why*, when we endeavor to obtain the area of circle by the *method of exhaustions*, using the straight line as a *means of measure*, we can get the *approximate* area only, and *why* it is necessary to obtain accurate results to use the principle of the *curve*. Geometry, in its present state, is the science of the straight line, and the introduction of the principle of the curve into geometrical consideration would usher in a *distinct science*, but eminently useful in solving problems of curvilinear spaces and boundaries which were before unsolved, inasmuch as the *approximate* results only were given for them.

The *method of exhaustions* is applicable to rectilineal magnitudes, and its results are consistent with the principle of the straight line, because the straight line is adopted as a *means* and *standard* of measurement; but since the straight line and curve *do not coincide*, the principle and properties of the straight line are not applicable to curvilinear spaces or boundaries; hence, what is true in one case, becomes absurd in the other.

PROP. XVIII.—THEOR.—*The sections of a solid by parallel planes are similar figures.*

Let the prism MN be cut by the two parallel planes AD, FK; their sections with it are similar figures.

For (VI. 7, cor.) the sections have parallel sides (I. 15, cor.

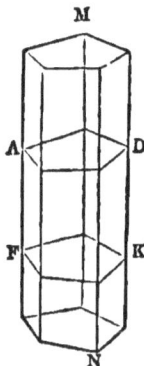

2). The figures AD, FK, therefore, have their sides similar, each to each. Their several angles are also (VI. 4) equal; for they are contained by straight lines which are parallel; and therefore the figures are similar.

Cor. 1. A section of a prism by a plane parallel to the base is equal and similar to the base.

Scho. 1. Since (VI. def. 24) a cylinder is described by the revolution of a rectangle about one of its sides, it is plain that any straight line in the rectangle perpendicular to the fixed line will describe a circle parallel to the base; and hence every section of a cylinder by a plane parallel to the base is a circle equal to the base.

Cor. 2. The section of a pyramid by a plane parallel to its base is a figure similar to the base.

Scho. 2. Since (VI. def. 21) a cone is described by the revolution of a right-angled triangle about one of its legs, it is plain that any straight line in the triangle perpendicular to the fixed leg will describe a circle parallel to the base; and the radius of that circle will be to the radius of the base as the altitude of the cone cut off to that of the whole cone.

Cor. 3. A section of a sphere by a plane is a circle.

Since the radii of the sphere are all equal, each of them being equal to the radius of the describing semicircle, it is plain that if the section pass through the center, it is a circle of the same radius as the sphere. But if the plane do not pass through the center, draw (VI. 5) a perpendicular to it from the center, and draw any number of radii of the sphere to the intersection of its surface with the plane. These radii, which are equal, are the hypothenuses of right-angled triangles, which have the perpendicular from the center as a common leg; and therefore (I. 24, cor. 2) their other legs are all equal; wherefore the section of the sphere by the plane is a circle, the center of which is the point in which the perpendicular cuts the plane.

Scho. 3. All the sections through the center are equal to one another, and are greater than the others. The former are therefore called *great circles*, the latter *small* or *less circles*.

Scho. 4. A straight line drawn through the center of a circle of the sphere perpendicular to its plane is a diameter of the sphere. The extremities of this diameter are called the *poles* of the circle. It is plain (I. 24, cor. 2) that chords drawn in the sphere from either pole of a circle to the circumference are all equal; and therefore (III. 16, cor. 3) that arcs of great circles between the pole and circumference are likewise equal.

Scho. 5. The pyramid or cone cut off from another pyramid or cone by a plane parallel to the base is similar (VI. defs. 8 and 27) to the whole pyramid or cone.

PROP. XIX.—THEOR.—*If the altitude of a parallelopiped, and the length and perpendicular breadth of its base be all*

*d:vided into parts equal to one another, the continued p oduct
of the number of parts in the three lines is the number of cubes
contained in the parallelopiped, each cube having the side of its
base equal to one of the parts.*

First, suppose the parallelopiped to be rectangular. Then
planes parallel to the base passing through the points of section
of the altitude will evidently divide the so'id into as many
equal solids as there are parts in the altitude; and each of
these partial solids will be composed of as many cubes as the
base contains squares, each equal to a base of one of the cubes.
But (I. 23, cor. 4) the number of these squares is the product
of the length and breadth of the base; and hence the entire
number of cubes will be equal to the product of the three
dimensions, the length, breadth, and altitude.

If the base be not rectangular, its area (I. 23, cor. 5) will be
the product of its length and perpendicular breadth; and it is
evident that the product of this by the altitude will be the
number of cubes as before.

Lastly, if the insisting lines be not perpendicular to the base,
still the oblique parallelopiped is equal (VI. 15, and cor. 1) to
a rectangular one of the same altitude; and therefore the number of cubes will be found as before, by multiplying the area
of the base by the altitude.

Cor. 1. Hence it is evident that the *volume,* or *numerical
solid content,* or, as it is also called, the *solidity,* of a parallelopiped is the product of its altitude and the area of its base
both expressed in numbers; and it is plain that the same
holds in regard to any prism whatever, and also in regard to
cylinders.

Cor. 2. The content of a pyramid or cone is found by multiplying the area of the base by the altitude, and taking a third
of the product. For (VI. 17) a pyramid is a third part of a
prism, and (VI. 17) a cone a third part of a cylinder, of the
same base and altitude.

An easy method of computing the content of a truncated
pyramid or cone, that is, the frustum which remains when a
part is cut from the top by a plane parallel to the base, may be
thus investigated by the help of algebra. The solid cut off is
(VI. 18, scho. 3) similar to the whole; and therefore the areas

of their bases will be proportional to the squares of their corresponding dimensions, and consequently to the squares of their altitudes. Hence putting V to denote the volume or content of the frustum, H and B the altitudes and base of the whole solid, and h and b those of the solid cut off, if we put qH^2 to denote B, since B $:b::$ H^2 $:h^2$, or B $:b::$ qH^2 $:qh^2$, we shall have (IV. 2, cor. 1) $b=qh^2$; and therefore (VI. 19, cor. 2) the contents of the whole cone and the part cut off are equal respectively to $\frac{1}{3}q$H^3 and $\frac{1}{3}qh^3$; wherefore V$=\frac{1}{3}q($H$^3-h^3)$, or, by resolving the second member into factors, V$=\frac{1}{3}q($H$^2+$Hh $+h^2)$ $($H$-h)=\frac{1}{3}(q$H^2+qH$h+qh^2)$ $($H$-h)$. Now qH^2 is equal to B, qh^2 to b, qHh to a mean proportional between them, and H$-h$ to the height of the frustum. Hence, *to find the content of a truncated pyramid or cone, add together the areas of its two bases and a mean proportional between them, multiply the sum by the height of the frustum, and divide the product by 3.*

This admits of convenient modifications in particular cases. Thus, if the bases be squares of which S and s are sides, and if a be the altitude of the frustum, we shall have

$$V=\tfrac{1}{3}a(S^2+Ss+s^2)=\tfrac{1}{3}a(3Ss+S^2-2Ss+s^2) \ ;$$
$$\text{or, } V=\tfrac{1}{3}a\{3Ss+(S-s)^2\}=a\{Ss+\tfrac{1}{3}(S-s)^2\}.$$

Hence. *to find the content of the frustum of a square pyramid, to the rectangle under the sides of its bases add a third of the square of their difference, and multiply the sum by the height.* It would be shown in like manner (V. 25, scho. and VI. 17, cor. 2), that if R and r be the radii of the bases of the frustum of a cone, and a its altitude,

$$V=\tfrac{1}{3}a\{Rr+\tfrac{1}{3}(R-r)^2\}.$$

Solidity of cylinder, $3\times R^2\times a$.

Solidity of cone, $R^2\times a$.

Solidity of sphere, is $4R^3$.

Solidity of spherical sector, is $2R^2\times a$.

Solidity of spherical segment, when it has two bases, is

$$\tfrac{3}{2} R^2+r^2)\times a+\tfrac{1}{2}a^3 \ ;$$

and when it has but one base,

$$\tfrac{3}{2}R^2\times a+\tfrac{1}{2}a^3.$$

Cor. 3. The content of a polyhedron may be found by dividing it into pyramids, and adding together their contents. The division into pyramids may be made either by planes passing through the vertex of one of the solid angles, or by planes passing through a point within the body.

PROP. XX.—THEOR.—*The surfaces of two similar polyhedrons may be divided into the same number of similar triangles similarly situated.*

This follows immediately from the definition (VI. def. 8) of similar bodies bounded by planes, if the sides or faces of the polyhedron be triangles; and any face in the one, and the corresponding face in the other, which are not triangles, are yet similar, and may be divided (I. 20) into the same number of similar triangles similarly situated.

Cor. Hence it would be shown, as in the fourteenth proposition of the fifth book, that the surfaces of the polyhedrons are proportional to any two of their similar triangles; and therefore they are to one another in the duplicate ratio of the homologous sides of those triangles, that is, of the edges or intersections of the similar planes. Hence also the surfaces are proportional (V. 14, cor. 2) to the squares of the edges.

PROP. XXI.—THEOR.—*Triangular pyramids are similar, if two faces in one of them be similar to two faces in the other, each to each, and their inclinations equal.*

Let ABC, abc be the bases, and D, d the vertices of two triangular pyramids, in which ABC, DBC are respectively similar to abc, dbc, and the inclination of ABC, DBC equal to that of abc, dbc; the pyramids are similar.

To demonstrate this, it is sufficient to show that the triangles ABD, ACD are similar to abd, acd, for then the solid angles (VI. 9, cor. 3) will be equal, each to each, and (VI. def. 8) the pyramids similar. Since the plane angles at B and b are equal, the inclinations of ABC, DBC, and of abc, dbc, are (VI. 9, cor. 2) equal; therefore ABD, abd are equal. Then (hyp.) DB : BC :: db : bc, and BC : BA :: bc :

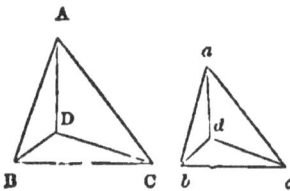

ba ; whence, *ex æquo,* DB : BA : : *db : ba ;* and therefore (V. 6) the triangles ABD, *abd* are equiangular, and consequently similar; and it would be proved in the same manner that ACD, *acd* are similar. Therefore (VI. def. 8) the pyramids are similar.

Cor. Hence triangular pyramids are similar, if three faces of one of them be respectively similar to three faces of the other.

In the triangular pyramids ABCD, *abcd* (see the preceding figure), let the faces ABC, ABD, DBC be similar to *abc, abd, dbc,* each to each; the pyramids are similar.

For (V. def. 1) AD : DB :: *ad : db,* and DB : DC :: *db : dc ;* whence, *ex æquo,* AD : DC :: *ad : dc.* Also DC : CB :: *dc : cb,* and CB : CA :: *cb : ca ;* whence, *ex æquo,* DC : CA :: *dc : ca ;* and therefore (V. 5) the triangles ADC, *adc* are equiangular, and (VI. 9, cor. 3, and def. 8) the pyramids are similar.

PROP. XXII.—THEOR.—*Similar polyhedrons may be divided into the same number of triangular pyramids, similar, each to each, and similarly situated.*

Let ABCDEFG and *abcdefg* be similar polyhedrons, having the solid angles equal which are marked with the corresponding large and small letters; they may be divided into the same number of similar triangular pyramids similarly situated.

The surfaces of the polygons may be divided (VI. 20) into the same number of similar triangles, similarly situated; then planes passing through any two corresponding solid angles, A, *a,* and through the sides of all these triangles, except those

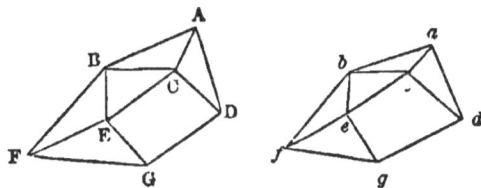

forming the solid angles, A, *a,* will divide the polyhedrons into triangular pyramids, similar to one another, and similarly situated.

The pyramids thus formed have each one solid angle at the common vertex A or *a ;* and these solid angles may be of three classes: 1*st,* those which have two of their faces coinciding with faces of one of the polyhedrons; 2*d,* those which have only one face coinciding ; and 3*d,* those which lie wholly within the solid angle A or *a.* Now those of the first kind in one of the polyhedrons are similar to the corresponding ones in the other, by the corollary to the twenty-first proposition of this book ; and those of the second kind by the twenty-first From the polyhedrons take two of these similar pyramids, and the remaining bodies will be similar, as the boundaries common to them and the pyramids are (VI. 21, and cor.) similar triangles ; and their other boundaries are similar, being faces of the proposed polyhedrons. Also the solid angles of the remaining bodies are equal, as some of them are angles of the primitive polyhedrons, and the rest are either trihedral angles which are contained by equal plane angles, or may be divided into such. From these remaining bodies other similar triangular pyramids may be taken in a similar manner, and the process may be continued till only two similar triangular pyramids remain ; and thus the polyhedrons are resolved into the same number of similar triangular pyramids.

Prop. XXIII.—Prob.—*To find the diameter of a given sphere.*

Let A be any point in the surface of the given sphere, and take any three points B, C, D at equal distances from A. Describe the triangle *bcd* having *bc* equal to the distance or chord BC, *cd* equal to CD, and *bd* to BD. Find *e* the center of the

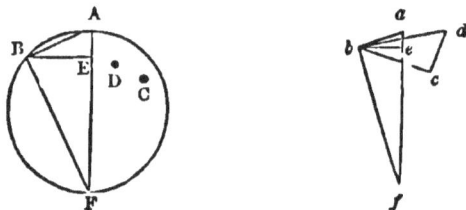

circle described about *bcd,* and join *be;* draw *aef* perpendicular to *be,* and make *ba* equal to BA ; draw *bf* perpendicular to *ba,* and *af* is equal to the diameter of the sphere.

Conceive a circle to be described through BCD, and E to be its center; that circle will evidently be the section of the sphere by a plane through B, C, D; and it will be equal to the circle described about *bcd*. Conceive the diameter AEF to be drawn, and BA, BE, BF to be joined. Then, in the right-angled triangles ABE, *abe*, the sides AB, BE are respectively equal to *ab*, *be*, and therefore (I. 24, cor. 2) the angles A, *a* are equal.

Again: in the right-angled triangles ABF, *abf*, the angles A, *a* are equal, and also the sides AB, *ab;* hence (I. 14) the sides AF, *af* are equal; that is, *af* is equal to the diameter of the sphere.

Prop. XXIV.—Theor.—*The angle of a spherical triangle is the angle formed by the tangents of the arcs forming the spherical angle, and is measured by the arc of a great circle described from the vertex as a pole, and intercepted by the sides, produced if necessary.*

Let BAC be a spherical angle formed by the arcs AB and AC, then it is the same as the angle EAD formed by the tangents EA and DA, and is measured by the arc of a great circle intercepted by the arcs AB and AC, produced if necessary. The tangents AE and AD are both perpendicular to the common diameter AH (III. 12), and being in the same planes as the arcs AB and AC, form an angle EAD equal to the spherical angle BAC.

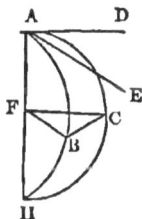

Again: the radii FB and FC of the great circle described from the vertex as a pole, being in the same planes as the arcs AB and AC, and perpendicular to AH, are parallel with AE and AD respectively, hence the angle EAD is equal to the angle BFC. But the angle BFC is measured by the arc BC (I. def. 19); therefore, also (I. ax. 1), the spherical angle BAC is measured by the arc BC.

Scho. The angles of spherical triangles may be compared together by means of the arcs of great circles described from their vertices as poles and included between the arcs forming the angles, and it is easy to make a spherical angle equal to a given angle.

Cor. 1. If from the vertices of the three angles of a spherical triangle as poles, arcs be described forming a spherical triangle, then the vertices of the angles of this second triangle will be respectively poles of the sides of the first, and each angle will be measured reciprocally by a semicircumference *less* the side of the other triangle opposite to the angle.

Because A, B, and C are poles respectively of the arcs FE, ED, and DF, the distances of the poles from the extremities of their respective arcs are, in each case, a quadrant; hence the

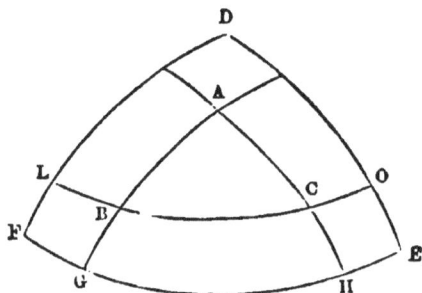

extremities of the arcs FE, ED, and DF are respectively removed the length of a quadrant from the extremities of the arcs AB, BC, and AC; therefore the extremities of the former arcs are the poles of the latter arcs, each to each.

Since A is the pole of the arc GH, the angle BAC is measured by that arc (VI. 24), and F being the pole of AH, FH is a quadrant, and E being the pole of AG, GE is a quadrant; hence FH + GE $\bigcirc$ semicircumference; but FH + GE $\bigcirc$ FE + GH, or the arc GH, which measures the angle BAC, is equivalent to a semicircumference *less* the arc FE. In like manner, the angle ABC can be shown to be measured by a semicircumference less the arc DE, and the angle ACB to be measured by a semicircumference less the arc DF. And, reciprocally, the angle FDE is measured by the arc LO. But LO + BC $\bigcirc$ LC + BO $\bigcirc$ semicircumference; hence LO $\bigcirc$ semicircumference minus BC; and a similar condition can be shown for the other angles of the triangle FED.

Cor. 2. As each angle of a spherical triangle is less than two right angles, the three angles are less than six right angles. And as the sum of the sides of a spherical triangle is less than the circumference of a great circle, and the angles being measured (VI. 24, cor. 1) by three semicircumferences less the three sides of the polar triangle, taking away the sides, we have the remainder greater than one semicircumference—or the three angles greater than two right angles. Hence the angles of a spherical triangle vary between two right angles and six right angles, without reaching either limit; therefore two angles given can not determine the third.

12

END OF BOOK SIXTH.

THE ELEMENTS OF PLANE TRIGONOMETRY.

1. TRIGONOMETRY is the practical application of geometrical principles for the investigation of ratios of the sides of triangles in connection with the magnitudes of their angles. For *perspicuity*, the vertex of the angle is placed in the center of a circle, and the arc of the circumference intercepted by the sides containing the angle is used as a *measure* of the angle (I. def. 19). Let a straight line be supposed to move around a fixed point; it will make with a stationary line angles which will vary as the line is moved, and when it has passed around until it coincides with the stationary line from which it is supposed to have started, it will have gone over the magnitude of four right angles (I. 9, cor.), the extremity of the movable line will trace the circumference of a circle, and the successive arcs intercepted between the movable and stationary lines will give the magnitude of the angles (I. def. 19).

2. For the purposes of calculation, a right angle is divided into an arbitrary number of equal parts; each one of these parts is subdivided into other equal parts, and each part of this subdivision undergoes a second subdivision of equal parts, and when particular nicety and precision are desired, there is a third subdivision of equal parts. Thus, the first division of a right angle is into degrees. Among the English mathematicians, a right angle has ninety degrees, which division is derived from Greek works, and has great antiquity, being used by the remotest ancient mathematicians and astronomers of whom we have any account. Among some modern French mathematicians, a right angle is divided into one hundred degrees, which centesimal division is continued throughout all the various subdivisions. But in the Greek division, which is more generally used on account of the great facility with which 360 can be subdivided, each degree has sixty equal

parts called *minutes*, each minute has sixty equal parts called *seconds*, and each second is sometimes subdivided in *decimal* parts, and thus the most extreme minuteness can be obtained.

3. The symbols for abbreviation in the expression of the value of angles are as follows: °, ′, ″; thus, 60°, 15′, 25″ are read, sixty degrees, fifteen minutes, and twenty-five seconds.

4. The reason why the right angle is assumed for division is because that angle preserves an *inversion* between every angle less than a right angle and its *complement* (I. def. 20), and a *similarity* between every angle greater than a right angle and its *supplement* (I. def. 20); thus, in the first case, the functions of the angle are *inverted* in respect to its complement, and in the latter case the functions are *the same* in respect to its supplement, as will more readily be seen by the seventh definition and following corollaries.

5. The straight line drawn from one extremity of an arc, perpendicular to the diameter passing through the other extremity, is the *sine* of the angle measured by it; and the part of that diameter intercepted between the sine and the arc is the *versed sine* of the angle which it measures.

6. If a straight line touch a circle at one extremity of an arc, the part of it intercepted between that extremity and the diameter produced, which passes through the other, is the *tangent* of the angle which it measures; and the straight line drawn from the center to the remote extremity of the tangent is the *secant* of the angle.

7. The *cosine* of an angle is the sine of its complement. In like manner, the *coversed sine*, *cotangent*, and *cosecant* of an angle are respectively the versed sine, tangent, and secant of its complement.

The sine, versed sine, tangent, and secant may be denoted by the abbreviated expressions, *sin*, *versin* (or *vs.*), *tan*, and *sec ;* and the cosine, coversed sine, cotangent, and cosecant, by *cos*, *coversin* (or *covs*), *cotan*, and *cosec.* For the sake of simplicity, the radius of the circle employed for comparing different angles is generally taken in investigations as unity; when this is not done, it is denoted by its initial letter R.

The sides of a triangle are often conveniently denoted by the small letters corresponding to the capital ones placed at the opposite angles. Thus, a denotes the side opposite to the angle A, etc. To prevent ambiguity, we may read A, B, C; *angle* A, *angle* B, *angle* C; while a, b, c may be called *side a, side b, side c.*

To illustrate the foregoing definitions, let C be the center of a circle, and AB, DE two diameters perpendicular to each other. Through any point F in the circumference draw the diameter FL; draw FG perpendicular to AB, and FI to DE; through A draw AH perpendicular to AB, and therefore (III. 8, cor.) touching the circle in A ; and let it meet LF produced in H ; and, lastly, draw DK perpendicular to DE, meeting FL

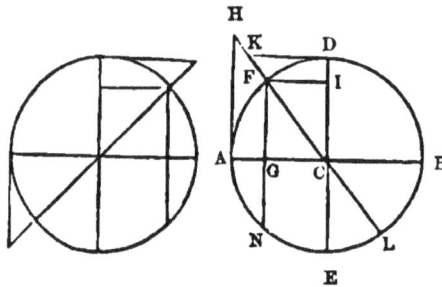

produced in K. Then the arc AF contains the same number of degrees, etc., as the angle ACF; and FG is the sine of this angle; FI, or its equal CG, the cosine; AG the versed sine, and DI the coversed sine ; AH the tangent, and CH the secant; DK the cotangent, and CK the cosecant.

From these definitions we derive the following corollaries :

Cor. 1. The sine of an angle ACF is half the chord of double the arc measuring it. For if FG be produced to meet the circumference in N, FN is bisected (III. 2) in G, and (III. 17) the arc FAN in A.

Cor. 2. The sine of the right angle ACD is the radius CD.

Cor. 3. If AF be half of AD, and consequently ACF half a right angle, the tangent AH is equal to the radius. For A being a right angle, H must be half a right angle, and (I. 1, cor. 2) AH equal to AC.

Cor. 4. Put the angle$=$A, and the radius$=1$. Then (I. 24, cor. 1) $FG^2 + CG^2 = CF^2$; that is, $\sin^2 A + \cos^2 A = 1$. In like manner, we find from the right-angled triangles CAH, CDK, that $CH^2 = CA^2 + AH^2$, and $CK^2 = CD^2 + DK^2$; that is, $\sec^2 A = 1 + \tan^2 A$, and $\csc^2 A = 1 + \cot^2 A$.

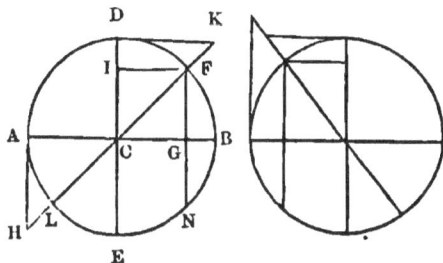

Cor. 5. In the similar triangles CGF, CAH, CG : CF, or CA :: CA : CH; that is, the cosine of an angle is to the radius as the radius to its secant. Hence also (V. 9, cor.) CG.CH $= CA^2$; that is, $\cos A \sec A = 1$. It would be found in like manner from the triangles CIF, CDK, that $\sin A \csc A = 1$, CI being equal to the sine FG.

Cor. 6. In the same triangles CGF, CAH, the cosine CG is to the sine GF as the radius CA to the tangent AH; whence (V. 8) $\cos A \tan A = \sin A$. The triangles CIF, CDK give in like manner $\sin A \cot A = \cos A$.

Cor. 7. The radius is a mean proportional between the tangent of an angle and its cotangent. For the triangles CAH, CDK are similar; and therefore HA : AC :: CD, or CA : DK. Hence (V. 9, cor.) $\tan A \cot A = 1$.

Cor. 8. The sine of an angle, and the sine of its supplement are equal. So likewise are their cosines, tangents, cotangents, secants, and cosecants.

Let ACF be an angle, FG, AH its sine and tangent, and CG, DK its cosine and cotangent. Make the angle BCM equal to ACF; draw the perpendicular MO; and produce MC both ways to meet HA, KD produced in P and Q. Then (I. def. 20, and I. 9) the angles BCM, ACM, or ACF, ACM are supplements of each other; as are also the arcs BM, AM, or AF, AM, since (III. 16) BM, AF are equal. Now the triangles CGF, COM are equiangular, and have the sides CF, CM equal;

therefore (I. 14) MO is equal to FG, and CO to CG; and MO, FG are the sines of ACM, ACF, and CO, CG their cosines. Again: the triangles ACP, ACH are equiangular, and have AC common; therefore (I. 14) AP is equal to AH, and CP to CH; and AP, AH are the tangents, and CP, CH the secants of ACM, ACF. In like manner it would be proved, by means of the triangles CDQ, CDK, that DQ, the cotangent of ACM, is equal to DK, the cotangent of ACF, and that their cosecants CQ, CK are equal.

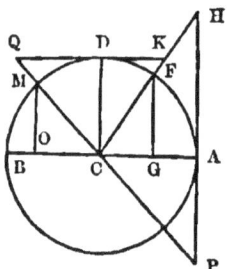

PROP. I.—THEOR.—*In a right-angled triangle the hypothenuse is to either of the legs as the radius to the sine of the angle opposite to that leg, or to the cosine of the adjacent angle; (2) either of the legs is to the other as the radius to the tangent of the angle opposite to the latter; and (3) either of the legs is to the hypothenuse as the radius to the secant of the contained angle.*

Let ABC be a triangle, right-angled at C; then (1) $c : b :: R : \sin B$, or $\cos A$; (2) $a : b :: R : \tan B$; and (3) $a : c :: R : \sec B$.

From B as center, with any radius, describe an arc cutting AB, BC in D, E; and through D, E draw (I. 8 and 7) DF, EG perpendicular to BC. Then (TRIG. defs. 5 and 6) FD, EG, and BG are respectively the sine, tangent, and secant of the angle B; and, since C is a right angle, A and B (TRIG. def. 4) are complements of each other; and therefore (TRIG. def. 7) $\sin B = \cos A$. Again: since the angle B is common to the triangles ABC, DBF, GBE, and the angles at C, F, E right angles, these triangles (I. 20, cor. 5) are equiangular.

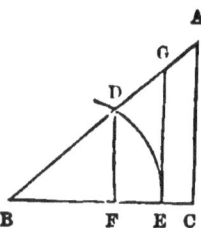

Hence (V. 3) in the triangles ABC, DBF,

BA : AC :: BD : DF; that is, $c : b :: R : \sin B$, or $\cos A$.

Again: (V. 3) in the triangles ABC, GBE,

BC : CA :: BE : EG; that is, $a : b :: $ R $:$ tan B; and

BC : BA :: BE : BG; that is, $a : c :: $ R $:$ sec B.

Cor. Hence (V. 10, cor.) R$b = c$ sin B$= c$ cos A; that is, *the product of either leg and the radius is equal to the product of the hypothenuse and the sine of the angle opposite to that leg, or of the hypothenuse and the cosine of the adjacent angle.* When R$= 1$, this becomes simply $b = c$ sin B$= c$ cos A. Again: $b = a$ tan B and $c = a$ sec B.

PROP. II.—THEOR.—*The sides of a plane triangle are proportional to the sines of the opposite angles.*

Let ABC be any triangle; then $a : b :: $ sin A $:$ sin B; $a : c :: $ sin A $:$ sin C; and $b : c :: $ sin B $:$ sin C.

Draw AD perpendicular to BC; then AD is a leg of each of

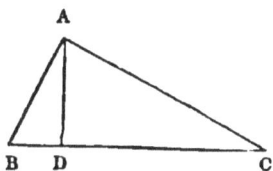

the right-angled triangles ADB, ADC; and therefore (TRIG. 1, cor.) R.AD$=$AB sin B, and R.AD$=$AC sin C. Hence (I. ax. 1) AB sin B$=$ AC sin C, or c sin B$= b$ sin C; whence (V. 10, cor.) $b : c :: $ sin B $:$ sin C; and, by drawing perpendiculars from B and C to the opposite sides, it would be proved in a similar manner that $a : c :: $ sin A $:$ sin C, and $a : b :: $ sin A $:$ sin B.

Cor. From B as center with BA as radius, describe an arc AD; and from C as center, with an equal radius, describe an arc EF. Draw AG, EH perpendicular to BC; these (TRIG. def. 5) are respectively the sines of B and C to equal radii. Then the triangles AGC, EHC are equiangular, the angle at C being common, and the angles at G and H right angles. Hence (V. 3) CA : AG :: CE, or (const.) AB : EH; and, alternately, CA : AB :: AG : EH; that is, $b : c :: $ sin B $:$ sin C.

The demonstration is simplified by taking, as here, one of the sides, AB or AC, as radius. This, however, is not essential, as arcs may be described from B and C as centers, with equal radii of *any* magnitude, and their sines, and a perpendicular from A to BC being drawn, the proof will be readily obtained.

Scho. From one of the foregoing analogies we have, by inversion, $c : b :: $ sin C $:$ sin B. If C be a right angle, this

(Trig. def. cor. 2) becomes $c : b :: $ R $: $ sin B, as in Prop. I. The first part, therefore, of that proposition is a particular case of this one.

Prop. III.—Theor.—*The sum of any two sides of a triangle is to their difference as the tangent of half the sum of the angle opposite to those sides is to the tangent of half their difference.*

Let ABC be a triangle, a, b any two of its sides, of which a is the greater, and A, B the angles opposite to them; then $a+b : a-b :: $ tan $\frac{1}{2}$(A+B) : tan $\frac{1}{2}$(A—B).

From C as center, with the greater side a as radius, describe the circle DBE, cutting AC produced in D and E, and BA produced in F; join BD, BE, CF; and draw EG parallel to AB, meeting DB produced in G.

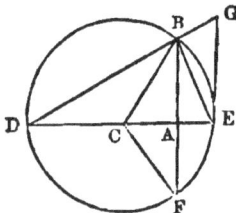

Then because DC and CE are each equal to a, DA is equal to $a+b$, and AE to $a-b$.

Also (I. 20) the exterior angle DCB is equal to A+B; and DEB, which is at the circumference, is (III. 10) half of DCB, which is at the center; therefore DEB=$\frac{1}{2}$(A+B).

Again (I. 1, cor. 1): the angle F is equal to B; and (I. 20) in the triangle ACF, the exterior angle A=ACF+F=ACF+B; and consequently, ACF=A—B; and (III. 10) ABE, or (I. 16) its equal, BEG=$\frac{1}{2}$(A—B).

Now, since (III. 11) EBD, being in a semicircle, is a right angle, as also (I. 9) EBG; if a circle were described from E as center, with EB as radius, DBG (III. 8, cor.) would touch it, and (Trig. def. 6) DB would be the tangent of DEB, and BG of BEG; and therefore DB, BG will evidently be proportional to the tangents of those angles to any other radius.

Or strictly, EB : BD :: 1 : tan DEB (Trig. 1) and (inver.) BD : EB :: tan DEB : 1. Also (Trig. 1) EB : BG :: 1 : tan BEG. Hence, *ex æquo*, BD : BG :: tan DEB : tan BEG.

Lastly, since BA (const.) is parallel to GE, we have (V. 2)
DA : AE :: DB : BG; that is,
$$a+b : a-b :: \text{tan } \tfrac{1}{2}(A+B) : \text{tan } \tfrac{1}{2}(A-B).$$

PROP. IV—THEOR.—*In a plane triangle, the cosine of half the difference of any two angles is to the cosine of half their sum, as the sum of the opposite sides to the third side; and* (2) *the sine of half the difference of any two angles is to the sine of half their sum, as the difference of the opposite sides to the third side.*

Let ABC (see the last proposition) be any plane triangle;

$$\text{then } \cos\tfrac{1}{2}(A-B) : \cos\tfrac{1}{2}(A+B) :: a+b : c;$$
$$\text{and } \sin\tfrac{1}{2}(A-B) : \sin\tfrac{1}{2}(A+B) :: a-b : c.$$

For it was shown in the preceding proposition, that $BED=\tfrac{1}{2}(A+B)$, and $ABE=\tfrac{1}{2}(A-B)$; and since DBE is a right angle, DBA is the complement of ABE, and D of BED. But (TRIG. 2) in the triangle ABD, $\sin ABD : \sin D :: AD : AB$; that is, (TRIG. def. 7) $\cos\tfrac{1}{2}(A-B) : \cos\tfrac{1}{2}(A+B) :: a+b : c$. Again (TRIG. 2) : in the triangle ABE, $\sin ABE : \sin AEB :: AE : AB$; that is, $\sin\tfrac{1}{2}(A-B) : \sin\tfrac{1}{2}(A+B) :: a-b : c$.

PROP. V.—THEOR.—*In any plane triangle the sum of the segments of the base made by a perpendicular from the vertex, is to the sum of the other sides as the difference of those sides to the difference of the segments.*

Let ABC be a triangle, and AD a perpendicular from the vertex to the base; the sum of the segments BD, DC is to the sum of the sides AB, AC, as the difference of AB, AC to the difference of BD, DC.

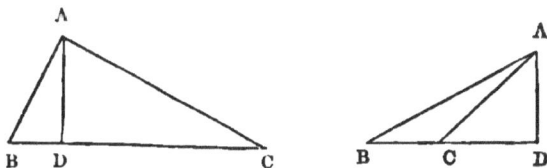

For (II. 5, cor. 4) the rectangle under the sum and difference of AB, AC is equivalent to the rectangle under the sum and difference of BD, DC; and therefore (V. 10, cor.) the sum of BD, DC is to the sum of AB, AC, as the difference of AB, AC to the difference of BD, DC.

Scho. If the perpendicular fall within the triangle, the segments make up the base, and their difference is less than the base; but if the perpendicular fall without the triangle, as it does (second fig.) when one of the angles at the base is obtuse, the base is the difference of the segments, and their sum is greater than the base.

PROP. VI.—THEOR.—*The rectangle under two sides of a triangle is to the rectangle under the excesses of half the perimeter above those sides, as the square of the radius to the square of the sine of half the contained angle.*

Let ABC be a triangle, and let $s=\frac{1}{2}(a+b+c)$; then $bc : (s-b)(s-c) :: R^2 : \sin^2\frac{1}{2}A$.

Produce the less side AC through C, making AD equal to AB; join BD; and draw AE, CF perpendicular, and CG parallel to BD; then (I. 24, cor. 2) AE bisects BD and the angle A. Now (II. 5, cor. 4) the rectangle under the sum and difference of BC, CD is equivalent to the rectangle under the sum and difference of BF, FD, that is, under BD and twice EF; therefore the rectangle under half the sum and half the difference of BC, CD is equivalent to the rectangle BE.EF. But (TRIG. 1)

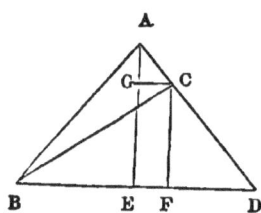

$$AB : BE :: R : \sin\tfrac{1}{2}A, \text{ and}$$
$$AC : CG \text{ or } EF :: R : \sin\tfrac{1}{2}A; \text{ whence (IV. 15)}$$
$$AC.AB : BE.EF, \text{ or } \tfrac{1}{2}(BC+CD).\tfrac{1}{2}(BC-CD) :: R^2 : \sin^2\tfrac{1}{2}A;$$
$$\text{or, } bc : \tfrac{1}{2}(a+c-b).\tfrac{1}{2}(a+b-c) :: R^2 : \sin\tfrac{1}{2}A,$$

because $CD=c-b$: and let $2s=a+b+c$; then, $s-a=\frac{1}{2}(b+c-a)$; $s-b=\frac{1}{2}(a+c-b)$; and $s-c=\frac{1}{2}(a+b-c)$; then we have $bc : (s-b)(s-c) :: R^2 : \sin^2\frac{1}{2}A$.

Cor. Hence, taking $R=1$ dividing the product of the means by the first extreme, and extracting the square root, we find

$$\sin\tfrac{1}{2}A=\sqrt{\frac{(s-b)(s-c)}{bc}},$$ and it is plain that we should find in a

similar manner, $\sin\frac{1}{2}B=\sqrt{\dfrac{(s-a)(s-c)}{ac}}$, and $\sin\frac{1}{2}C=$

$\sqrt{\dfrac{(s-a)(s-b)}{ab}}$.

PROP. VII.—THEOR.—*The rectangle under two sides of a triangle is to the rectangle under half the perimeter and its excess above the third side, as the square of the radius to the square of the cosine of half the angle contained by the two sides.*

Let ABC be a triangle, and let $s=\frac{1}{2}(a+b+c)$; then $bc : s$ $(s-a) :: R^2 : \cos^2 \frac{1}{2} A$.

Produce the less side CA, through A, making AD equal to AB; join BD; draw AE, CF perpendicular, and AG parallel to BD. Then BD (I. 24, cor. 2) is bisected in E; and the angle BAC being (I. 20) equal to the two equal angles D and ABD, each of them is equal to half the angle BAC, that is, half the angle A in the triangle ABC; and (I. 16) GAC is equal to D. Now, it would be shown, as in the preceding proposition, that the rectangle under half the sum and half the difference of DC, CB is equivalent to the rectangle BE.EF. But (TRIG. 1) AB : BE :: R : cos ABE, or cos $\frac{1}{2}$ A; and AC : AG, or EF :: R : cos GAC, or cos $\frac{1}{2}$ A; whence (IV. 15),

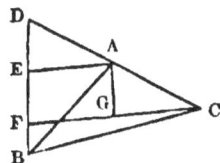

AC.AB : BE.EF, or $\frac{1}{2}$ (DC+CB). $\frac{1}{2}$ (DC—CB) :: R^2 : $\cos^2\frac{1}{2}$A;

or $bc : \frac{1}{2}(a+b+c). \frac{1}{2}(b+c-a) :: R^2 : \cos^2\frac{1}{2}$ A,

because DC$=b+c$. If $2s$ be the same as last proposition :

$$bc : s(s-a) :: R^2 : \cos^2\frac{1}{2} A.$$

Cor. 1. Hence we find, as in the corollary to the preceding proposition, that

$$\cos\tfrac{1}{2}A=\sqrt{\frac{s(s-a)}{bc}};$$ and it would be proved in a similar manner, that

$$\cos\tfrac{1}{2}B=\sqrt{\frac{s(s-b)}{ac}}, \text{ and } \cos\tfrac{1}{2}C=\sqrt{\frac{s(s-c)}{ab}}.$$

Cor. 2. From the sixth corollary to the definitions of trigonometry, it is plain, when the radius is unity, if the sine of an angle be divided by its cosine, the quotient is its tangent. Hence, by dividing the expression for the sine of $\frac{1}{2}$ A in the corollary to the preceding proposition, by the value of its cosine in the last corollary, we obtain $\tan\frac{1}{2}A=\sqrt{\dfrac{(s-b)(s-c)}{s(s-a)}};$

and we should obviously find in a similar manner, that

$$\tan \tfrac{1}{2}B = \sqrt{\frac{(s-a)\ (s-c)}{s\ (s-b)}}, \text{ and } \tan \tfrac{1}{2}C = \sqrt{\frac{(s-a)\ (s-b)}{s\ (s-c)}}.$$

Cor. 3. By dividing the values of $\tan \tfrac{1}{2}B$, $\tan \tfrac{1}{2}C$, in the preceding corollary, each by that of $\tan \tfrac{1}{2}A$, we obtain

$$\frac{\tan \tfrac{1}{2}B}{\tan \tfrac{1}{2}A} = \frac{s-a}{s-b}, \text{ and } \frac{\tan \tfrac{1}{2}C}{\tan \tfrac{1}{2}A} = \frac{s-a}{s-c}.$$

PROP. VIII.—PROB.—*Given the radius of a circle, and the cosine of an angle, less than a right angle; to compute the cosine of half the angle.*

Let CD, the cosine of ACB, less than a right angle, be given; it is required to compute the cosine of its half.

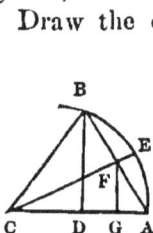

Draw the chord AB, and perpendicular to it draw CFE; draw also FG parallel to BB; then (III. 2, and TRIG. defs. 5 and 7) AF or FB is the sine, and CF the cosine of ACE the half of ACB. Also (V. 2), DG is equal to GA, since BF is equal to FA, and DA=2DG; to each of these add 2CD; then CA+CD=2CG, and consequently $CG = \tfrac{1}{2}(CA + CD)$; or, if the radius be taken as unity, and the angle ACB be denoted by A, $CG = \tfrac{1}{2}(1 + \cos A)$. Again, in the similar triangles ACF, CFG, AC : CF :: CF : CG; whence (V. 10, cor. 2) $CF^2 = AC.CG$; that is, $\cos 2 \tfrac{1}{2} A = \tfrac{1}{2}(1 + \cos A)$. Hence, *to compute* $\cos \tfrac{1}{2} A$, *add 1 to* $\cos A$, *take half the sum, and extract the square root.*

Scho. This proposition and the next afford means by which trigonometrical tables can be computed.

PROP. IX.—THEOR.—*If A and B be any two angles,* R : $\cos B :: \sin A : \tfrac{1}{2} \sin (A-B) + \tfrac{1}{2} \sin (A+B)$.

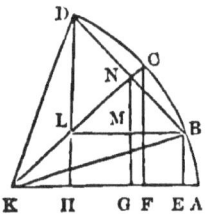

Make AKC equal to A, and BKC, CKD each equal to B; draw BE, CF, DH, the sines of AKB, AKC, AKD; join BD, and through the center K draw KNC; then KN is evidently the cosine of BKC. Draw also BML, NMG parallel to AK, DH. Now, in the similar triangles DLB, NMB, since DB is double of NB, DL (V. 3) is double of NM; to

DL add LII, BE, and to 2NM add what is equivalent, 2MG; then DH+BE=2NG; wherefore NG is equal to half the sum of BE and DH, that is, to $\frac{1}{2}$ sin (A—B)+$\frac{1}{2}$ sin (A+B). Again, in the similar triangles CFK, NGK, we have (V. 3, and alternately) CK : NK :: CF : NG; that is, R : cos B :: sin A : $\frac{1}{2}$ sin (A—B)+$\frac{1}{2}$ sin (A+B).

Cor. 1. Hence, if R=1, we have, by doubling the second and fourth terms, and by taking the products of the extremes and means, sin (A—B)+ sin (A+B)=2 cos B sin A; whence, sin (A+B)=2 cos B sin A—sin (A—B).

Cor. 2. If B=A, the last expression becomes simply sin 2 A =2 sin A cos A.

TRIGONOMETRICAL FORMULÆ.

The lines hitherto considered may be computed for every conceivable angle, and they will each undergo a change of value when the angle passes through the gradations of magnitude, hence they are the *functions* of the angle, a term implying the connection between two varying quantities, that the value of the one changes with the value of the other, and they receive their values from the ratios or proportions arising from them and the angle. We have considered the numerical values only of these functions, and the angles from which they were deduced were all less than 180 degrees, which relate to plane angles and triangles. And we propose now to explain the processes for computing the unknown parts of rectilinear triangles, also the nature and properties of the angular functions, together with the methods of deducing all the formulæ which express relations between them.

When two diameters are drawn perpendicularly to each other, they divide the circle into four equal parts called *quadrants*, which are first, second, third, and fourth quadrants, going from right to left, and the functions have certain *algebraic* values depending upon the particular quadrant in which the angle is. For instance, all the lines estimated from AC *upward* are *positive*, and from CA

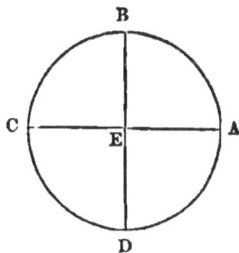

downward are *negative ;* from DB to *the right* are *positive,* and from DB to *the left* are *negative.* In the formulæ the algebraic signs + and — are used, the former denoting positive, and the latter negative.

In the diagram it will be seen that the functions are all positive in the first quadrant AEB; that the sine, cosecant, and versed sine are positive, and the others negative in the second quadrant; that the tangent, cotangent, and versed sine are positive, and the others negative in the third quadrant; that the cosine, secant, and versed sine are positive, and the others negative in the fourth quadrant. Hence we can arrange them in the following table:

	First Quad.	Second Q.	Third Q.	Fourth Q.
Sine,	+	+	—	—
Cosine,	+	—	—	+
Tangent,	+	—	+	—
Cotangent,	+	—	+	—
Secant,	+	—	—	+
Cosecant,	+	+	—	—
Versed sine,	+	+	+	+

It is convenient to give different signs to the angles also. If we suppose the angles to be estimated from left to right, they are negative, and the sign of the angle will affect the sign of its sine, but those of its cosine remain the same.

From corollaries fourth, fifth, sixth, and seventh we can deduce the following equations, when A denotes the angle and the radius is unity:

$$\operatorname{Sin}^2 A + \cos^2 A = 1 \qquad (1)$$

$$\operatorname{Sec}^2 A = 1 + \tan^2 A \qquad (2)$$

$$\operatorname{Cosec}^2 A = 1 + \cot^2 A \qquad (3)$$

$$\operatorname{Tan} A = \frac{\sin A}{\cos A} \qquad (4)$$

$$\operatorname{Cot} A = \frac{\cos A}{\sin A} \qquad (5)$$

$$\operatorname{Tan} A \times \cot A = 1 \qquad (6)$$

$$\operatorname{Sec} A = \frac{1}{\cos A} \qquad (7)$$

$$\text{Cosec } A = \frac{1}{\sin A} \qquad (8)$$

$$\text{Ver. } \sin A = 1 - \cos A \qquad (9)$$

By the first proposition we have, in a right-angled triangle, radius : cos of either acute angle :: hyp. : side adjacent.
Hence, in the following diagram:

$CB \bumpeq CA \cos C$; and $DB \bumpeq DA \cos D$,
or $CD \bumpeq CA \cos C - DA \cos D$.

Dividing both members of the equation by CD, we have—

$1 \bumpeq \dfrac{CA}{CD} \cos C - \dfrac{DA}{CD} \cos D$; hence (TRIG. 2)

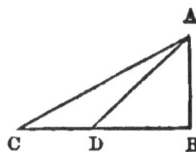

$\dfrac{CA}{CD} \bumpeq \dfrac{\sin D}{\sin A}$ and $\dfrac{DA}{CD} \bumpeq \dfrac{\sin C}{\sin A}$; hence, $1 \bumpeq \dfrac{\sin D}{\sin A} \cos C - \dfrac{\sin C}{\sin A}$
cos D, or $\sin A \bumpeq \sin D \cos C - \sin C \cos D$; but the angle A is the difference of the angles ADB and ACB (I. 20); hence $\sin (D - C) \bumpeq \sin D \cos C - \sin C \cos D$.

Thus, from the first and second propositions, by easy processes, we derive the formula for the sine of the difference of two angles, which is expressed in the following manner: *The sine of the difference of any two angles is equivalent to the sine of the first into the cosine of the second, minus the cosine of the first into the sine of the second.**

The formula for the sine of the sum of two angles can be derived from the preceding by substituting the negative for the positive value of the second angle, and bearing in mind that in estimating an angle from *left to right*, the algebraic sign of its sine is changed, and we get—*The sine of the sum of any two angles is equivalent to the sine of the first into the cosine of the second, plus the cosine of the first into the sine of the second.*

The formula for the cosine of the sum of two angles can be derived from the preceding, by substituting the *trigonometrical* values of the functions when the sum of the angles is greater than a right angle, and remembering that the sine of an angle becomes the cosine of its complement, and we get—*The cosine*

---

* The pupil would be much instructed by converting this and the following expressions into their equivalent algebraic formulæ.

*of the sum of two angles is equivalent to the cosine of the first
into the cosine of the second, minus the sine of the first into the
sine of the second.*

By similar substitutions in the second formula, we get the
formula for the cosine of the difference of two angles, expressed
as follows: *The cosine of the difference of two angles is equivalent to the cosine of the first into the cosine of the second, plus
the sine of the first into the sine of the second.*

The other corresponding formulæ are obtained by substituting the respective equations for the trigonometrical *functions*
derived from the fourth, fifth, sixth, and seventh corollaries for
the values of the various functions; for instance, to derive the
formula for the tangent of the sum of two angles, we substi-

tute, $\tan A = \dfrac{\sin A}{\cos A}$ in the second and third formulæ, getting

$$\tan (A+B) = \frac{\sin (A+B)}{\cos (A+B)} = \frac{\sin A \cos B + \cos A \sin B}{\cos A \cos B - \sin A \sin B},$$ and re-

ducing, we have—*The tangent of the sum of two angles is
equivalent to the tangent of the first, plus the tangent of the second, divided by the square of the radius, minus the tangent of
the first into the tangent of the second.* And in a similar way
we get the other formulæ for the various functions.

When the angle and radius are known, we can get the formula for the sine of double the angle by making the two angles
equal in the second formula, and we have—*The sine of twice
an angle is equivalent to twice the sine of the angle into the
cosine of the angle.*

In similar manner we derive the formula for the cosine of
double the angle, and substituting the equation $\sin^2 A = 1 -
\cos^2 A$ in the third formula, we get—*The cosine of twice an
angle is equivalent to twice the square of the cosine of the angle, minus the square of the radius.*

Thus by substitution of the several equations in the respective formulæ, we can derive the other functions of double the
angle when the angle and radius are known.

From the formula for the cosine of double an angle, we can
get by substitutions and reductions the formula for the sine of
half an angle, and expressed as follows—*The sine of half an
angle is equivalent to the square root of half the difference of*

the radius and cosine of the angle. In a similar way we obtain—*The cosine of half an angle is equivalent to the square root of half the sum of the radius and cosine of the angle;* and—*The tangent of half an angle is equivalent to the sine of the angle divided by the sum of the radius and cosine of the angle;* and so on for the other functions.

By adding and subtracting the various formulæ already mentioned, we obtain a great number of consequences which are useful; it will suffice to consider a few of them. From the first four we obtain—*The sine of the sum of two angles added to the sine of the difference of the same angles is equivalent to twice the sine of the first into the cosine of the second. The sine of the sum of two angles diminished by the sine of the difference of the same angles is equivalent to twice the sine of the second into the cosine of the first. The cosine of the sum of two angles increased by the cosine of the difference of the same angles is equivalent to twice the cosine of the first into the cosine of the second. The cosine of the difference of two angles diminished by the cosine of the sum of the same angles is equivalent to twice the sine of the first into the sine of the second.* These are very useful, because they change the products of sines, cosines, and other functions from *superficial* into *linear* sines, cosines, etc.

By the substitution of algebraic symbols into the preceding equations, we obtain certain analogies which give algebraic expressions to so many theorems, as follows:

Sum of sines : Dif. of sines :: tan of half Sum : tan of half Dif.
Sum of sines : Sum of cos :: tan of half Sum : radius.
Sum of sines : Dif. of cos :: cot of half Dif. : radius.
Dif. of sines : Sum of cos :: tan of half Dif. : radius.
Dif. of sines : Dif. of cos :: cot of half Sum : radius.
Sum of cos : Dif. of cos :: cot of half Sum : tan of half Dif.
Sum of sines : sine of Sum :: cos of half Dif. : cos of half Sum.
Dif. of sines : sine of Sum :: Sine of half Dif. : sine of half Sum.

INVESTIGATIONS OF THE METHODS OF COMPUTING TABLES OF
SINES, TANGENTS, AND SECANTS.

From what has been shown in relation to the previous formulæ, it will be noticed that they all proceed from the first—and we derive the first from the first and second propositions, namely, from the analogies that in a right-angled triangle: hypothenuse : radius :: one of the legs : sine of opposite angle; hence, when we know the length of the hypothenuse, leg, and radius, we can determine the sine of the angle opposite the leg. Then (TRIG. def. 7, cor. 4) $\cos^2 A = 1 - \sin^2 A$; that is, the square of the radius, minus the square of the sine of the angle, is equivalent to the square of the cosine of the angle; then, square root of the difference between the squares of the radius and sine is equivalent to the cosine of the angle; and the other functions are derived from the equations resulting from the fourth, fifth, sixth, and seventh corollaries of the definitions.

Now, the chord of 60 degrees (III. 25, cor. 4) is equal to radius, and when radius is unity, the cosine of 30° is 0.5; hence, sine of $30° = \sqrt{(1-\cos^2 30°)} = \sqrt{.75}$. Bisecting this angle we get from the formula of cosine of half an angle, $\cos 15° = \frac{1}{2}\sqrt{1+\cos 30°}$, and seventeen such bisections give $\cos 1'' = .999999299$; hence, $\sin 1''$ can be obtained.

Some mathematicians divide 3.14159265358979, etc., into as many equal parts as there are seconds or minutes in 180°, and express the value of the sine of one second or minute by one of these equal parts, contending that the sine, chord, or arc of so small an angle differ very imperceptibly from each other; when this quantity is used, the cosine of one second becomes .999999957. It will thus be seen that the cosines obtained by these methods differ very little from each other, and for ordinary purposes either will give results sufficiently accurate; but when the greatest exactness is desired, the first method should be used, because the sine of an angle is a straight line and can never coincide with the arc which measures (I. def. 19) the angle, however so small the angle be reduced, and (V. 25, scho.) 3.1415926, etc., is the *approximate* relation of the diameter and circumference. Hence, the first method is a pure deduction from geometrical and trigonometrical principles. When the angle is a right angle, the sine and cosine of the angle are equal, and the equations of the tangent and cotangent will give the values of those functions; and when the angle exceed a right angle, the formula for the tangent of the sum of two angles can be used, making the right angle one of the angles and the excess the other angle.

When the converging series are used, any angle less than 90° is expressed by x, and the formulæ for the functions are:

$$\text{Sin } x = x - \frac{x^3}{1.2.3} + \frac{x^5}{1.2.3.4.5} - \frac{x^7}{1.2.3.4.5.6.7} + \text{etc.}$$

$$\text{Cos } x = 1 - \frac{x^2}{1.2} + \frac{x^4}{1.2.3.4} - \frac{x^6}{1.2.3.4.5.6} + \text{etc.}$$

$$\text{Tan } x = x + \frac{x^3}{3} + \frac{2x^5}{3.5} + \frac{17x^7}{3^2.5.7} + \frac{62x^9}{3^2.5.7.9} + \text{etc.}$$

$$\text{Cot } x = \frac{1}{x} - \frac{x}{3} - \frac{x^3}{3^2.5} - \frac{2x^5}{3^3.5.7} - \frac{x^7}{3^5 5^2 7} - \text{etc.}$$

$$\text{Sec } x = 1 + \frac{x^2}{1.2} + \frac{5x^4}{1.2.3.4} + \frac{61x^6}{1.2.3.4.5.6} + \text{etc.}$$

$$\text{Cosec } x = \frac{1}{x} + \frac{x}{1.2.3} + \frac{7x^2}{3.4.5.6} + \frac{31x^5}{3.4.5.6 \cdot 27} + \text{etc.}$$

Now, when the base of the Napierian logarithms is used, $e = 2.7182818$, and the following formulæ will give the sine and cosine of any angle x, from which the other functions can be obtained:

$$\text{Sin } x = \frac{e^{x\sqrt{-1}} - e^{-x\sqrt{-1}}}{2\sqrt{-1}}; \text{ and } \cos x = \frac{e^{x\sqrt{-1}} + e^{-x\sqrt{-1}}}{2}.$$

Having obtained the sine and cosine of any angle by either of the foregoing formulæ, we can get the sine of twice the angle by the consequences from adding and subtracting the first four formulæ, page 193, from which we derive—

$$2 \cos x \times \sin x - \sin 0 = \sin 2x,$$
$$2 \cos x \times \sin 2x - \sin x = \sin 3x,$$
$$2 \cos x \times \sin 3x - \sin 2x = \sin 4x,$$
$$\text{etc.,} \quad \text{etc.,} \quad \text{etc.,} \quad \text{etc.}$$

Or by multiplying the first two formulæ, page 191, and substituting the value of the square of the cosine, we can determine new formulæ for further computation after having found the sines of x and $2x$.

Sin $(x+2x)$ sin $(x-2x) = \sin^2 x - \sin^2 2x$; hence, sin $(x+2x)$

sin $(x-2x)=($sin $x+$sin $2x)$ (sin $x-$sin $2x)$; or, sin $(x-2x)$: sin $x-$sin $2x$ :: sin $x+$sin $2x$: sin $(x+2x)$; applying this proportion, we have,

Sin x : sin $2x-$sin x :: sin $2x+$sin x : sin $3x$.
Sin $2x$: sin $3x-$sin x :: sin $3x+$sin x : sin $4x$.
Sin $3x$: sin $4x-$sin x :: sin $4x+$sin x : sin $5x$.
     etc.,   etc.,   etc.,   etc.

These last formulæ will give the *natural* functions of the angles, but to avoid the operations of multiplication and division, and employ the simpler operations of addition and subtraction, tables are constructed giving the logarithmic values of the several functions of the angles.

As the sine and cosine of an angle are the legs of a right-angled triangle, and the hypothenuse is the radius of the arc which measures the angle, for the convenience of logarithms the hypothenuse or radius is considered as 10,000,000,000, and its logarithm is 10.

The sines, cosines, tangents, and cotangents are the only functions put in the tables, as the other functions are easily found from them.

TRIGONOMETRICAL PROBLEMS.

The principles which have been thus established, enable us to solve all the elementary cases of plane trigonometry. Now, of the three sides and three angles of a triangle, some three, and those not the three angles, must be given to determine the triangle (I. 14, scho.), and the resolution of plane triangles may therefore be reduced to the three following cases:

I. When a side and the opposite angle, and either another side or another angle are given;

II. When two sides and the contained angle are given;

III. When the three sides are given

The FIRST CASE is solved on the principle (TRIG. 2) that the sides are proportional to the sines of the opposite angles. Thus, if A, B, a be given, add A and B together, and take the sum from 180°; the remainder (I. 20) is C. Then b and c will be found by the following analogies : sin A : sin B : : a : b; and sin A : sin C : : a : c.

If, again, a, b, A be given, we compute B by the analogy, a : b : : sin A : sin B. Then, C is found by subtracting the sum of A and B from 180°, and c by the analogy, sin A : sin C : : a : c.

When in this case two unequal sides, and the angle opposite to the less, are given, the angle opposite to the greater (TRIG. defs. cor. 8) may be either that which is found in the table of sines or its supplement; and thus the problem admits of two solutions (I. 3, case 4).

If in this case one of the angles be a right angle, the solution is rather easier; as, by the second corollary to the definitions, the sine of that angle is equal to the radius. The same conclusion may also be obtained by means of the first proposition.

To exemplify the solution of this case,* let $a=13$ yards, $b=15$ yards, and A$=53°$ 8',—to resolve the triangle; and the operation by means of logarithms will be as follows:

As a	13		1.113943
: b	15		1.176091
: : sin A	53°	8'	9.903108
: sin B	67°	23',	9.965256
	or 112°	37'	
As sin A			9.903108
: sin C	59°	29'	9.935246
: :	a		1.113943
:	c	14	1.146081
As sin A			9.903108
: sin C	14°	15'	9.391206
: :	a		1.113943
:	c	4	0.602041

---

* For logarithmic computations the pupil is referred to the Tables now in preparation by Prof. Docharty, of the College of the City of New York, or the Tables computed by Prof. Davies, of the United States

In these operations, to find the fourth term, the second and third terms are added together, and the first is taken from the sum. This may be done very easily, in a single operation, by adding the figures of the second and third terms successively to what remains after taking the right-hand figure of the first term from 10, and each of the rest from 9, and rejecting 10 from the final result. Thus, in the first operation, we have 8 and 1 are 9, and 7 are 16; then 1 and 9 are 10, and 5 are 15, etc. It is still easier, however, when the quantity to be subtracted is a sine, to use the cosecant, and when it is a cosine, to use the secant, each diminished by 10, and then to add all the terms together. The reason of this is evident from the nature of logarithms, and from the fifth corollary to the definitions of Trigonometry. In like manner, when the number to be subtracted is a tangent, or cotangent, we may use in the former case a cotangent—in the latter, a tangent, subtracting in each case 10, either at first or afterward.

This example evidently belongs to the doubtful case; and hence we have two values for each of the quantities B, C, and c; and therefore two analogies are requisite for finding the values of c.

The SECOND CASE is solved by means of the third and fourth propositions. Thus, if a, b, C be taken, take C from 180°, and (I. 20) the remainder is $A + B$. Take the half of this, and then, by the third proposition, as $a+b : a-b :: \tan \frac{1}{2}(A+B) : \tan \frac{1}{2}(A-B)$. This analogy gives half the difference of A and B; and (II. 12, scho.) by adding this and $\frac{1}{2}(A+B)$ together, A, the greater angle, is obtained, while B is found by taking $\frac{1}{2}(A-B)$ from $\frac{1}{2}(A+B)$. The remaining side will be calculated (TRIG. 4) by means of either of the following analogies, and by employing both, an easy verification of the process is obtained :

as $\cos \frac{1}{2}(A-B) : \cos \frac{1}{2}(A+B) :: a+b : c$; and
$\sin \frac{1}{2}(A-B) : \sin \frac{1}{2}(A+B) :: a-b : c$.

When the given angle C is a right angle, the solution is most easily effected by means of the first proposition of Trig-

Military Academy, or those of Prof. Loomis, of Yale College, New Haven, Connecticut.

onometry; the oblique angles being obtained by the analogy, $a : b :: R : \tan B$, or $\cot A$; and the hypothenuse either by the analogy, $R : \sec B :: a : c$, or $\sin A : R :: a : c$.

As an example,

As $a+b$	99.98	1.999913
$: a-b$	14.78	1.169674
$:: \tan \frac{1}{2} (A+B)$	61° 37'½	10.267498
$: \tan \frac{1}{2} (A-B)$	15° 18'½	9.437259

Hence $A = 76° 56'$, and $B = 46° 19'$

As $\cos \frac{1}{2} (A-B)$	15° 18'½	9.984311
$: \cos \frac{1}{2} (A+B)$		9.676913
$:: a+b$		1.999913
$: c$	49.26	1.692515

As $\sin \frac{1}{2} (A-B)$		9.421626
$: \sin \frac{1}{2} (A+B)$		9.944411
$::\quad a-b$		1.169674
$:\quad c$	49.26	1.692459

Half the difference of A and B is here taken as 15° 18'½. When determined accurately, however, it is found to be 15° 18' 23''. Hence the cause of the slight difference in the logarithm of c, as obtained by the two different analogies. It is plain that after A and B are computed, c might also be found by means of the first case, by either of the analogies; $\sin A : \sin C :: a : c$, and $\sin B : \sin C :: b : c$. The foregoing method, however, is much preferable.

The THIRD CASE may be solved by means of the fifth, sixth, or seventh proposition. Thus, assuming a (see the figures for the fifth proposition) as base, we have a to $b+c$ as $b-c$, or $c-b$ to a fourth proportional. If this be less than BC, it is the difference of the segments BD, DC, in the first figure; and if half of it and half of the base be added together, the sum will be the greater segment, while the less will be found by taking half that proportional from half the base. If the fourth pro-

portional be greater than the base, it is the sum of the segments in the second figure, and, as before, the segments are the
sum and difference of half the proportional and half the base.
Then, by resolving, by the first case, the two right-angled triangles ADB, ADC, in which there are given the hypothenuses
AB, AC, and the legs BD, CD, the angles B and ACD will be
obtained, which, in the first figure, are two of the required angles; while in the second, the angle C is the supplement of
ACD.

Again: by adding the three sides together, and taking half
the sum, the value of s is obtained; and, in applying the sixth
proposition, the sides containing the required angle are to be
separately taken from s; but, in applying the seventh, only
the side opposite to the required angle is to be subtracted;
while if all the three sides be subtracted successively, another
mode of solution is furnished by the second and third corollaries to the seventh proposition. This last method is preferable
to any of the others, when it is necessary to determine all the
angles; and if they be all computed by means of it, the cor-
rectness of the operation is ascertained by trying whether their
sum is 180°.

To exemplify the last of these methods, let $a=679$, $b=537$,
and $c=429$; to compute the angles.

Here, by adding the three sides together, we obtain 1645,
the half of which, 822.5, is s. Then, by taking from the three
sides successively, we find $s-a=143.5$, $s-b=285.5$, and $s-c=393.5$. The rest of the operation, the subtraction in the first
part of which may be performed in the manner pointed out in
the example for the first case, is as follows:

s	822.5	2.915136	$\Big\}$ subt.
$s-a$	143.5	2.156852	
$s-b$	285.5	2.455606	
$s-c$	393.5	2.594945	
	2)	19.978563	
tan $\frac{1}{2}$A 44° 17$\frac{1}{2}'$		9.989281	
A$=88°$ 35$'$			

tan ½ A	9.989281	} add
log (s — a)	2.156852	

	12.146133	} subt.
log (s — b)	2.455606	

tan ½ B 26° 7′ ½	9.690527	
B = 52° 15′		

	12.146133	} subt.
log (s — c)	2.594945	

tan ½ C 19° 35′	9.551188	
C = 39° 10′		

In the first part of this operation, the halving of the logarithm serves for the extraction of the square root. The remainder of the work consists in adding together tan ½ A and log (s — a), and subtracting log (s — b) and log (s — c) successively from the sum. This method of solution is remarkably easy, requiring for the entire operation only four logarithms to be taken from the tables; and affording at the same time a most satisfactory verification by the addition of the three angles, when found. The preparatory part of the process also admits of an easy verification, as the sum of the three remainders s — a, s — b, s — c is equal to the half sum.

Prob. I.—*Let it be required to find the height of an* accessible *object AB, standing on a horizontal plane.*

On the horizontal plane take a station C, and measure with a line, a chain, or any such instrument, the distance CB to the base of the object; and with a quadrant, a theodolite, or other angular instrument, measure the angle BCA, called the *angle of elevation.* Then, since B is a right angle, the height AB will be found (Trig. 1) by the analogy, R : tan C :: CB : BA.

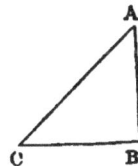

This gives the height of A above CB, the horizontal line passing through the eye of the observer; and therefore to find the entire height, AB must be increased by the height of his eye above the base or the object. The like addition must be made in every problem of this kind, when the angle of elevation above the horizontal line is given.

PROB. II.—*To find the height of an object* AB, *standing on a horizontal plane, but* inaccessible *on account of the unevenness of the ground near its base, or the intervention of obstacles.*

In a straight line passing through the base of the object take two stations C, D; and measure CD, and the two angles of elevation BCA, BDA. Then (I. 20) CAD is the difference of ACB, ADB; and (TRIG. 2) sin CAD : sin D :: DC : CA. Again (TRIG. 1), R : sin ACB :: AC : AB; whence AB will be found.

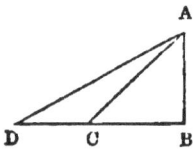

The computation will be rendered rather more easy by multiplying together (IV. 15) the terms of the two analogies, and dividing the third and fourth terms of the result by AC; as by this means we get the analogy R × sin CAD : sin D × sin BCA :: DC : AB. Hence, to find the logarithm of AB, to the logarithm of DC add the logarithmic sines of D and BCA, and from the sum take the sine of CAD and the radius.

PROB. III.— *To find the distance of two objects* A *and* B *on a horizontal plane.*

This may be effected in different ways according to circumstances.

1. A base AC may be taken, terminated at one of the objects. The angles A and C, and the side AC are then measured; and the required distance AB is found by the analogy, sin B : sin C :: AC : AB.

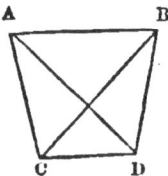

2. This method fails if the objects A and B be not visible from one another, as then the angle A can not be measured. In this case, a station C may be taken as before, from which both A and B are visible. Then, having measured the angle C, and the sides AC, BC, we compute the distance AB by means of the second case of trigonometry.

3. When from inequalities in the ground, or other causes, the preceding methods are inapplicable, the solution may be effected in the following manner: Measure a base CD, such that A and B are both visible from each of its extremities; measure, also, the two angles at C, and the two at D. Then, by the first case in trigonometry, we compute AC in the triangle ACD, and BC in the triangle BCD; from which, and from the contained angle ACB, AB is computed by means of the second case. The operation may be verified by computing AD, BD, by the first case, and thence AB by the second case.

PROB. IV.—*Let AFB be a great circle of the earth, supposed to be a sphere; E a point in the diameter BA produced, EF a straight line touching the circle in F, and ED a straight line in its plane, perpendicular to AB; it is required to compute the angle DEF, and the straight line EF.*

Draw the radius CF. Then, since (III. 12) CFE is a right angle, we have (hyp. and I. 20, cor. 3) $DEC = CEF + ECF$. Take away CEF, and there remains $DEF = ECF$. Now (III. 21) $EF^2 = BE.EA$. Hence EF will be found by adding AE to AB, multiplying the sum by EA, and extracting the square root. To find CE, add AE to the radius AC. Then (TRIG. 1) CE : EF :: R : sin ECF, or sin DEF; or CE : CF :: R : cos ECF, or cos DEF.

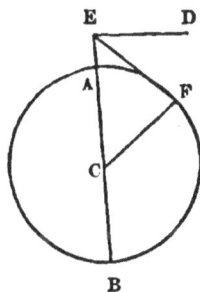

Scho. These examples have been selected from the "Elements of Plane Trigonometry," by Prof. Thomson, of the University of Glasgow, since they exhibit in the simplest manner the elementary principles of trigonometrical computation.

4*

THE ELEMENTS OF SPHERICAL TRIGONOMETRY.

1. SPHERICAL TRIGONOMETRY explains the processes of calculating the unknown parts of a spherical triangle when any three parts are given; and certain formulæ derived from Plane Trigonometry are employed to express the relations between the six parts of a spherical triangle.

2. A *spherical triangle* is that portion of the surface of a sphere bounded or contained by the arcs of three great circles intersecting each other; the spherical triangle being formed by three planes passing through the sphere, and intersecting each other, each angle of the triangle (VI. 24) is contained by the tangents of the sides at their point of intersection, and is measured by the arcs of great circles described from the vertices as poles, and limited by the sides of the triangle produced if necessary. Also, the angles of a spherical triangle vary (VI. 24, cor. 2) between *two* and *six* right angles.

3. The spherical angle being contained by the tangents of the sides at their point of intersection, the properties of the spherical triangle are explained by means of Plane Trigonometry, and its analogies are applied to *imaginary* rectilineal triangles, the sides of the spherical triangle being regarded *functions* of rectilineal angles having the sides of the spherical triangle as arcs *measuring* (I. def. 19) them. Spherical Trigonometry treats of the angles at the apex of a triangular pyramid; but Plane Trigonometry treats of *plane angles;* therefore Spherical Trigonometry treats of solid angles.

4. Let ABC be a spherical triangle, and H the center of the sphere; the angles of the triangle are equal to the angles included by the planes HAB, HAC, and HBC (VI. 24), which are the angles formed by the planes at the apex of a triangular pyramid, and the arcs AB, BC, and CA measure the angles on the planes at the apex of the pyramid AHB, BHC, AHC re-

spectively. And we can represent the side opposite the angle
A by a, the side opposite the
angle B by b, and the side opposite the angle C by c. On
the line HA take any point as
L, and draw perpendiculars as
FL, LG to HA. Then, GLF
will be equal to the angle A,
LFG equal to B, and FGL
equal to C; and the sides CB,
BA, and AC of the spherical triangle ABC will measure the
angles CHB, AHB, and AHC respectively; hence these angles are denoted by a, c, b.

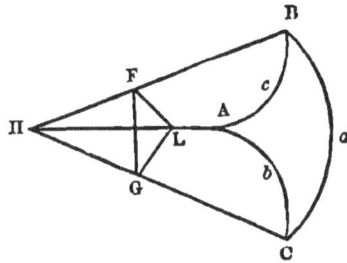

5. If FG be joined in the triangles FHG and FLG, we will
have (Plane Trig., 6 cor.),

$$\cos BHC = \cos a = \frac{HF^2 + HG^2 - FG^2}{2HF \times HG};$$

$$\cos FLG = \cos A = \frac{LF^2 + LG^2 - FG^2}{2LG \times LF}.$$

Reducing and subtracting second from the first, we will
have,

$$2 [\cos a (HF \times HG) - \cos A (LG \times LF)] = 2HL^2.$$

Dividing both members by 2 (HF × HG), we get,

$$\cos a - \cos A \frac{LG \times LF}{HF \times HG} = \frac{HL}{HF} \times \frac{HL}{HG}.$$

Since regular and similar polygons have their perimeters
proportionate to their apothems, and circles have their circumferences proportionate to their diameters (V. 14, cor. 3), the
sine of an angle is the ratio of the radius, or the hypothenuse
of a right-angled triangle to the perpendicular from the vertex
of the right angle to the hypothenuse.

Hence we get $\frac{LG}{HG} = \sin b$, $\frac{LF}{HF} = \sin c$, $\frac{HL}{HF} = \cos c$, $\frac{HL}{HG} = \cos b$.

Substituting and transposing, we derive the formulæ,

$$\left.\begin{array}{l}\cos a = \cos b \cos c + \sin b \sin c \cos A, \\ \cos b = \cos a \cos c + \sin a \sin c \cos B, \\ \cos c = \cos a \cos b + \sin a \sin b \cos C.\end{array}\right\} \quad (1)$$

Or, *The cosine of either side of a spherical triangle is equal to the product of the cosines of the other two sides* increased by *the product of their sines into the cosine of the angle included by them.*

The three formulæ show the relations between the *six* parts of a spherical triangle such, that if any *three* of them be given, the other *three* can be determined.

6. Then (VI. 24, cor. 1), if we denote the angles and sides of the spherical triangle polar to ABC, respectively, by A′, B′, C′, a′, b′, c′, we will have,

$$a' = 180° - A, \ b' = 180° - B, \ c' = 180° - C,$$
$$A' = 180° - a, \ B' = 180° - b, \ C' = 180° - c.$$

Since any of the formulæ (1) is applicable to polar spherical triangles, we have, after substituting the respective values and changing the signs of the terms, other formulæ :

$$\left. \begin{array}{l} \cos A = \sin B \sin C \cos a - \cos B \cos C, \\ \cos B = \sin A \sin C \cos b - \cos A \cos C, \\ \cos C = \sin A \sin B \cos c - \cos A \cos B. \end{array} \right\} \quad (2)$$

Or, *The cosine of either angle of a spherical triangle is equal to the product of the sines of the two other angles into the cosine of their included side, diminished by the product of the cosines of those angles.*

7. Transposing the first and second formulæ (1) we get,

$$\cos a - \cos b \cos c = \sin b \sin c \cos A,$$
$$\cos b - \cos a \cos c = \sin a \sin c \cos B;$$

respectively adding and subtracting, we get,

$$\cos a + \cos b - \cos c \ (\cos a + \cos b) = \sin c \ (\sin b \cos A + \sin a \cos B),$$
$$\cos a - \cos b + \cos c \ (\cos a - \cos b) = \sin c \ (\sin b \cos A - \sin a \cos B);$$

which can be put in the forms

$$(1 - \cos c) \ (\cos a + \cos b) = \sin c \ (\sin b \cos A + \sin a \cos B),$$
$$(1 + \cos c) \ (\cos a - \cos b) = \sin c \ (\sin b \cos A - \sin a \cos B);$$

multiplying these equations together, substituting for $(1 - \cos^2)$ its value $(\sin^2)$, and for $(\cos^2)$ its value $(1 - \sin^2)$, and dividing by $\sin^2 c$, we have,

$\cos^2 a - \cos^2 b = \sin^2 b - \sin^2 b \sin^2 A - \sin^2 a + \sin^2 a \sin^2 B;$

then, since $(1 - \sin^2 a) - (1 - \sin^2 b) = \sin^2 b - \sin^2 a$, we have,

$\cos^2 a - \cos^2 b = \sin^2 b - \sin^2 a$, and we get,

$\sin^2 b \sin^2 A = \sin^2 a \sin^2 B;$

extracting the square root, $\sin b \sin A = \sin a \sin B$, or

$\dfrac{\sin A}{\sin B} = \dfrac{\sin a}{\sin b}.$ And from first and third formulæ (1) we

derive $\dfrac{\sin A}{\sin C} = \dfrac{\sin a}{\sin c}.$ From the second and third formu-

læ we derive $\dfrac{\sin C}{\sin B} = \dfrac{\sin c}{\sin b}.$

$$\left.\begin{array}{c}\\[2em]\\[2em]\\\end{array}\right\} (3)$$

Or, *In every spherical triangle, the sines of the angles are to each other as the sines of their opposite sides.*

8. Taking the third formula (1), and substituting for ($\cos b$) its value as expressed in the second, and for ($\cos^2 a$) its value $(1 - \sin^2 a)$, and dividing by $\sin a$, we will have,

$$\cos c \sin a = \sin c \cos a \cos B + \sin b \cos C.$$

But (Spher. Trig. 7) we have $\sin b = \dfrac{\sin B \sin c}{\sin C};$ substituting for $\sin b$ its value, and dividing by $\sin c$, we get, $\dfrac{\cos c}{\sin c} \sin a = \cos a \cos B + \dfrac{\sin B \cos C}{\sin C}.$

$$\text{But } \frac{\cos}{\sin} = \cot.$$

Hence we can derive, by similar processes,

$$\left.\begin{array}{l}
\cot a \, \sin b = \cot A \, \sin C + \cos b \, \cos C,\\
\cot a \, \sin c = \cot A \, \sin B + \cos c \, \cos B,\\
\cot b \, \sin a = \cot B \, \sin C + \cos a \, \cos C,\\
\cot b \, \sin c = \cot B \, \sin A + \cos c \, \cos A,\\
\cot c \, \sin a = \cot C \, \sin B + \cos a \, \cos B,\\
\cot c \, \sin b = \cot C \, \sin A + \cos b \, \cos A.
\end{array}\right\} (4)$$

The formulæ (1) are the fundamental analogies of Spherical Trigonometry, from which all the others are derived, which others are more adapted for logarithmic computations.

THE SOLUTION OF RIGHT-ANGLED SPHERICAL TRIANGLES BY LOGARITHMS.

The following equations give the unknown parts of a right-angled spherical triangle when C is the right angle and any two other parts are known. There are six cases.

Let C be the right angle, and c be the hypothenuse.

Case 1. Given a and b to find c, A, and B;

$$\cos c = \cos a \cos b; \quad \tan A = \frac{\tan a}{\tan b}; \quad \tan B = \frac{\tan b}{\tan a}.$$

Case 2. Given c and a side b to find a, A, and B;

$$\cos a = \frac{\cos c}{\cos b}; \quad \cos A = \frac{\tan b}{\tan c}; \quad \sin B = \frac{\sin b}{\sin c}.$$

Case 3. Given a side a and opposite angle A to find others;

$$\sin b = \frac{\tan a}{\tan A}; \quad \sin C = \frac{\sin a}{\sin A}; \quad \sin B = \frac{\cos A}{\cos a}.$$

Both acute or both not acute.

Case 4. Given a side a and adjacent angle B to find others;

$\tan b = \sin a \tan B$; $\cot c = \cot a \cos B$; $\cos A = \cos a \sin B$.

Case 5. Given the hypothenuse c and an angle A to find others;

$\sin a = \sin c \sin A$; $\tan b = \tan c \cos A$; $\cot B = \cos c \tan A$.

Case 6. Given the oblique angles A and B to find others;

$$\cos a = \frac{\cos A}{\sin B}; \quad \cos b = \frac{\cos B}{\sin A}; \quad \cos c = \cot A \cot B.$$

Napier's *circular parts* are much the simplest method of resolving *right-angled spherical triangles;* they are the two sides about the right angle, the complements of the oblique angles, and the complement of the hypothenuse. Hence there are *five* circular parts; the right angle not being a circular part, is supposed not to separate the two sides adjacent to the right angle; therefore these sides are regarded adjacent to each other, so that when any two parts are given, their corresponding circular parts are also known, and these with the required part constitute the three parts under consideration; therefore these three parts will lie together, or one of them will be separated from both the others. Hence one part is known as the *middle part;* and when three parts are under consideration, the parts separated by the middle part are called the *adjacent parts ;* and the parts separated from the middle parts are called the *opposite parts.* Now, assume any part in

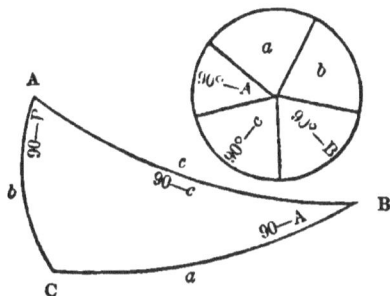

the diagram for the middle part, and using the formulæ (1) when the other parts are the opposite parts, and we get, *The sine of the middle part is equal to the product of the cosines of the opposite parts.* Then, assume again any part for the middle, and use the formulæ (2) when the other parts are the adjacent parts, and we get, *The sine of the middle part is equal to the product of the tangents of the adjacent parts.* Hence we derive the five following equations:

$$\sin a = \tan b \, \tan (90^\circ - B) = \cos (90^\circ - A) \, \cos (90^\circ - c),$$
$$\sin b = \tan a \, \tan (90^\circ - A) = \cos (90^\circ - B) \, \cos (90^\circ - c),$$
$$\sin (90^\circ - A) = \tan b \, \tan (90^\circ - c) = \cos (90^\circ - c) \, \cos a,$$
$$\sin (90^\circ - c) = \tan (90^\circ - A) \, \tan (90^\circ - B) = \cos a \, \cos b,$$
$$\sin (90^\circ - B) = \tan a \, \tan (90^\circ - c) = \cos b \, \cos (90^\circ - A).$$

The *analogies* of Napier are derived from the formulæ (1) by eliminating the cosines of any of the sides, reducing and changing to *linear* sines and cosines (Plane Trig. p. 193), and we have,

$$\cos \tfrac{1}{2} (a+b) : \cos \tfrac{1}{2} (a-b) :: \cot \tfrac{1}{2} C : \tan \tfrac{1}{2} (A+B),$$
$$\sin \tfrac{1}{2} (a+b) : \sin \tfrac{1}{2} (a-b) :: \cot \tfrac{1}{2} C : \tan \tfrac{1}{2} (A-B),$$
$$\cos \tfrac{1}{2} (a+c) : \cos \tfrac{1}{2} (a-c) :: \cot \tfrac{1}{2} B : \tan \tfrac{1}{2} (A+C),$$
$$\sin \tfrac{1}{2} (a+c) : \sin \tfrac{1}{2} (a-c) :: \cot \tfrac{1}{2} B : \tan \tfrac{1}{2} (A-C),$$
$$\cos \tfrac{1}{2} (b+c) : \cos \tfrac{1}{2} (b-c) :: \cot \tfrac{1}{2} A : \tan \tfrac{1}{2} (B+C),$$
$$\sin \tfrac{1}{2} (b+c) : \sin \tfrac{1}{2} (b-c) :: \cot \tfrac{1}{2} A : \tan \tfrac{1}{2} (B-C),$$

14

The same proportions applied to the triangle polar to ABC, with accents omitted, we have,

$$\cos \tfrac{1}{2}(A+B) : \cos \tfrac{1}{2}(A-B) :: \tan \tfrac{1}{2}c : \tan \tfrac{1}{2}(a+b),$$
$$\sin \tfrac{1}{2}(A+B) : \sin \tfrac{1}{2}(A-B) :: \tan \tfrac{1}{2}c : \tan \tfrac{1}{2}(a-b),$$
$$\cos \tfrac{1}{2}(A+C) : \cos \tfrac{1}{2}(A-C) :: \cot \tfrac{1}{2}b : \tan \tfrac{1}{2}(a+c),$$
$$\sin \tfrac{1}{2}(A+C) : \sin \tfrac{1}{2}(A-C) :: \cot \tfrac{1}{2}b : \tan \tfrac{1}{2}(a-c),$$
$$\cos \tfrac{1}{2}(B+C) : \cos \tfrac{1}{2}(B-C) :: \cot \tfrac{1}{2}a : \tan \tfrac{1}{2}(b+c),$$
$$\sin \tfrac{1}{2}(B+C) : \sin \tfrac{1}{2}(B-C) :: \cot \tfrac{1}{2}a : \tan \tfrac{1}{2}(b-c).$$

The same ambiguity that there is between plane triangles (I. 3, Fourth Case) exists also between spherical triangles, which may be avoided by remembering that every angle and side of a spherical triangle are each less than two right angles, and that the greater angle is opposite to the greater side, and the least angle is opposite to the least side; and conversely.

Quadrantal spherical triangles are such which have one side equal to ninety degrees; hence they can very easily be solved by formulæ for right-angled spherical triangles.

SOLUTION OF OBLIQUE-ANGLED SPHERICAL TRIANGLES BY LOGARITHMS.

Case 1. Given the three sides to find the angles.

Find $s = \tfrac{1}{2}(a+b+c)$; let $M = \sqrt{\dfrac{\sin (s-a) \sin (s-b) \sin (s-c)}{\sin s}}$;

then, $\tan \tfrac{1}{2}A = \dfrac{M}{\sin (s-a)}$; $\tan \tfrac{1}{2}B = \dfrac{M}{\sin (s-b)}$; $\tan \tfrac{1}{2}C = \dfrac{M}{\sin (s-c)}$.

Case 2. Given two sides a and b and the included angle C, to find others.

$\mathrm{Tan} \tfrac{1}{2}(A+B) = \dfrac{\cos \tfrac{1}{2}(a-b)}{\cos \tfrac{1}{2}(a+b)} \cot \tfrac{1}{2}C$; $\tan \tfrac{1}{2}A - B = \dfrac{\sin \tfrac{1}{2}(a-b)}{\sin \tfrac{1}{2}(a+b)} \cot \tfrac{1}{2}C$.

But $A = \tfrac{1}{2}(A+B) + \tfrac{1}{2}(A-B)$; $B = \tfrac{1}{2}(A+B) - \tfrac{1}{2}(A-B)$;

$\sin c = \sin a \dfrac{\sin C}{\sin A} = \sin b \dfrac{\sin C}{\sin B}$, or find an angle $\cot \varphi = \tan a \cos C$; $\cos c = \dfrac{\cos a \sin (b+\varphi)}{\sin \varphi}$.

Case 3. Given the sides a and b and an angle opposite to one of them, to find others.

Find $\cot \varphi = \tan b \cos A$, and $\tan \chi = \cos b \tan A$; then, $\sin$

$$(c+\varphi) = \frac{\cos a \sin \varphi}{\cos b}; \ \sin B = \sin A \frac{\sin b}{\sin a}; \ \sin (c+\chi) = \cot a \tan b$$

$\sin \chi$.

Case 4. Given the angles A and B and the included side, to find others.

Find $\tan \varphi = \cos c \tan A$, and $\tan \chi = \cos c \tan B$;

$$\text{then, } \tan a = \frac{\tan c \sin \varphi}{\sin (B+\varphi)}; \ \tan b = \frac{\tan c \sin \chi}{\sin (A+\chi)};$$

$$\cos C = \frac{\cos A \cos (B+\varphi)}{\cos \varphi} = \frac{\cos B \cos (A+\chi)}{\cos \chi}.$$

Case 5. Given A and B and a side opposite one of them, to find others.

Find $\tan \varphi = \tan a \cos B$; $\cot \chi = \cos a \tan B$;

$$\sin b = \sin a \frac{\sin B}{\sin A}; \ \sin (c-\varphi) = \cot A \tan B \sin \varphi,$$

$$\sin (C-\chi) = \frac{\cos A \sin \chi}{\cos B}.$$

Case 6. Given the three angles A, B, C, to find the sides. Take

$$S = \tfrac{1}{2}(A+B+C); \text{ and } N = \sqrt{\frac{-\cos S}{\cos (S-A) \cos (S-B) \cos (S-C)}};$$

$$\text{then, } \tan \tfrac{1}{2} a = N \cos (S-A);$$
$$\tan \tfrac{1}{2} b = N \cos (S-B);$$
$$\tan \tfrac{1}{2} c = N \cos (S-C).$$

THE SURFACE OF A SPHERICAL TRIANGLE.

Let ABC be a spherical triangle, AC= DF, BC=FE, ABC=DEF.

S=surface of ACB, s=surface of hemisphere — BHDC — AGEC — DCE=6 R²— (lune AHD — S)—(lune BGE— S)—(lune CDFE—S)

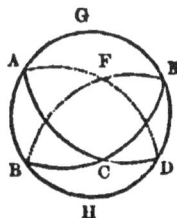

$$= 6 \text{ R}^2 \left(1 - \frac{A}{180} - \frac{B}{180} - \frac{C}{180}\right) + 3 \text{ S}.$$

$$S= \frac{A+B+C-180}{180} \; 6 \; R^2 = \frac{E}{180} \; 3 \; R^2, \; \therefore \; E = A+B+C-$$

180°, or *equivalent to the sum of the three angles above* 180°; hence, its *spherical excess* is sometimes taken as the measure of the triangle.

Or in terms of its sides, formula given by L'Huiller,

$$\tan \tfrac{1}{2} E = \sqrt{} \; [\tan \tfrac{1}{2} s \tan \tfrac{1}{2} (s-a) \tan \tfrac{1}{2} (s-b) \tan \tfrac{1}{2} (s-c)].$$

EXERCISES IN
ELEMENTARY GEOMETRY,

AND IN

PLANE AND SPHERICAL TRIGONOMETRY.*

DEFINITIONS.

1. *Lines, angles,* and *spaces* are said to be *given in magnitude,* when they are either exhibited, or when the method of finding them is known.

2. *Points, lines,* and *spaces* are said to be *given in position,* which have always the same situation, and which are actually exhibited or can be found.

3. A *circle* is said to be *given in magnitude* when its radius is given; and in *position,* when its center is given.

Magnitudes, instead of being said to be *given in magnitude,* or *given in position,* are often said simply to be *given,* when no ambiguity arises from the omission.

---

* For these Exercises I am indebted to Thomson's Euclid (Belfast), they being judiciously selected by that eminent writer, and their presentation here is a valuable acquisition to an American school text-book. I would gladly acknowledge my obligations for many propositions in this volume, but they being culled for more than two thousand years from the best writers on Geometry, and being so much modified by each succeeding age, that it is impossible at this day to attribute them to their rightful authors, and many of them being more or less contained in every work on the subject, they have become public property. What I have introduced myself will be well recognized by every student of Geometry, and my only apology is, the desire to advance the cause of Truth.

4. A *ratio* is said to be *given* when it is the same as that of two given magnitudes.

5. A *rectilineal figure* is said to be *given in species*, when its several angles and the ratios of its sides are given.

6. When a series of unequal magnitudes, unlimited in number, agree in certain relations, the greatest of them is called a *maximum ;* the least, a *minimum*.

Thus, of chords in a given circle, the diameter is the maximum; and of straight lines drawn to a given straight line, from a given point without it, the minimum line is the perpendicular.

7. A *line* which is such that any point whatever in it fulfills certain conditions, is called the *locus* of that point.

Scho. 1. Several instances of loci have already occurred in the preceding books.

1. Thus, it was stated in the fifth corollary to the fifteenth proposition of the first book, that all triangles on the same base, and between the same parallels, are equivalent in area; and hence, if only the base and area of a triangle be given, its vertex may be at any point in a straight line parallel to the base, and at a distance from it which may be determined by applying (II. 5, scho.) to half the base a parallelogram equivalent to the given area; and therefore the parallel is the locus of the vertex. Here the conditions fulfilled are, that straight lines drawn from any point in the parallel to the extremities of the given line, form with it a triangle having a given area.

2. It was stated in the first corollary to the eighteenth proposition of the third book, that all angles in the same segment of a circle are equal; and hence, if only the base and vertical angle of a triangle be given, the vertex may be at any point of the arc of a segment described on the base, in the manner pointed out in the nineteenth proposition of the third book; that arc, therefore, is the locus of the vertex.

3. It will be seen in the sixth proposition of these Exercises that straight lines drawn from any point whatever in the circumference of the circle ABGC to the points E, F, have the same ratio—that of EA to AF. Hence, therefore, if the base of a triangle, and the ratio of the sides be given, the locus of the vertex is the circumference of the circle described in the man-

ner pointed out in the corollary to this proposition; unless the ratio be that of equality, in which case the locus is evidently a perpendicular bisecting the straight line joining the points.

4. It follows likewise, from the fifth corollary to the twenty-fourth proposition of the first book, that when the base of a triangle and the difference of the squares of the sides are given, if the point D be found (II. 12, scho.) in the base BC, or its continuation, such that the difference of the squares of BD, CD is equivalent to the difference of the squares of the sides; and if through D a perpendicular be drawn to BC, straight lines drawn from any point of the perpendicular to B, C will have the difference of their squares equivalent to the given difference; and hence the perpendicular is the locus of the vertex, when the base and the difference of the squares of the sides are given.

5. It will appear in a similar manner from the corollary to the twelfth proposition of the second book, that if BC the base of a triangle, and the sum of the squares of the other sides AB, AC be given, the locus of the vertex is the circumference of a circle described from D, the middle point of the base as center, and with the radius DA. To find DA, take the diagonal of the square of BD as one leg of a right-angled triangle, and for its hypothenuse take the side of a square equivalent to the given sum of the squares of AB, AC; then the diagonal of the square described on half the remaining leg of that right-angled triangle will be the radius of DA. The proof of this is easy, depending on the third and fourth corollaries to the twenty-fourth proposition of the first book, and on the corollary to the twelfth proposition of the second book.

Scho. 2. In discovering loci, as well as in other investigations in geometry, the student is assisted by what is termed *geometrical analysis ;* of the nature of which it may be proper here to give some explanation.

Take this proposition : *If a chord of a given circle have one extremity given in position, and if a segment terminated at that extremity be taken on the chord, produced if necessary, such that the rectangle under the segment and chord may be equivalent to a given space ; the locus of the point of section is a straight line given in position.*

Let AB be the diameter of the circle and AC a chord of the circle.

If, in the proposition, instead of being informed that the locus is a straight line, we were required to find what the locus is, we might proceed in the following manner: Let D be any point in the required line, so that the rectangle AC.AD is equivalent to the given space; and having drawn the diameter AB, find E, so that the rectangle AB.AE may be equal to AC.AD, and therefore E a point in the required line; and join DE, BC. Then (V. 10, cor.) AB : AC : : AD : AE. Hence (V. 6) the triangles DAE, BAC, having the angle A common, are equiangular; and therefore AED is equal to ACB, which is a right angle. The point D is therefore in a perpendicular passing through E; and in the same manner it would be shown, that any other point in the required line is in the perpendicular; that is, the perpendicular is its locus.

The investigation just given is called the *analysis* of the proposition, while the solutions hitherto given are called the *synthesis* or *composition*. In analysis we commence by supposing that to be effected which is to be done, or that to be true which is to be proved; and, by a regular succession of consequences founded on that supposition, and on one another, we arrive at something which is known to be true, or which we know the means of effecting. Thus, in the second corollary to the seventeenth proposition of the sixth book, the conclusion obtained for the area of the circle is shown by the third corollary to be consistent with the proportion established by Archimedes between the cone, sphere, and cylinder, and also consistent with the geometrical truth in regard to the surfaces of the sphere and cylinder. Hence, analysis takes into consideration this consistence, and confirms the second corollary from its agreement with established truths of geometry.

Again: in the corollary to the twenty-fourth proposition of the fifth book, since circles are in proportion to the squares of their radii, the quadrant ACB is equivalent to the semicircle ADC, we have (I. ax. 3) the triangle ABC equivalent to the crescent ADC. Now, when we apply the conclusion derived by the second corollary to the seventeenth proposition of the sixth book to the above, we find a perfect agreement; taking

the circle as three times square of radius, we have quadrant
ACB equivalent to $\frac{3}{4}$ AB^2; hence (I. ax. 1) the semicircle ADC
is equivalent to $\frac{3}{4}$ AB^2. But (VI. 17, cor. 2) the segment AC
of the quadrant ACB is equivalent to $\frac{1}{4}$ AB^2; therefore (I. ax. 3)
we have the triangle ABC and the crescent ADC each equiva-
lent to $\frac{1}{2}$ AB^2, thus showing the agreement between the second
corollary to the seventeenth proposition of the sixth book, and
the established truth relating to the crescent or lune.

Also, we have (VI. 17, cor. 2) the hemisphere generated by
the quadrant BNP equivalent to the solid generated by the
trapezium BSNP, and we have the solid generated by the
figure BTNP common; therefore (I. ax. 3) the solid generated
by the segment BT is equivalent to the solid generated by the
figure TSN. Now, the solid generated by the segment BT is
a part of the hemisphere; hence its contents are computed by
the same radius as the hemisphere; the solid generated by the
figure TSN is a part of the solid generated by the trapezium
BSNP; hence its contents are also computed by the same
radius as the hemisphere. Therefore we obtain by analysis
the truth, *that when equivalent solids are generated by equivalent surfaces, the generating surfaces are up n equal radii*, a
truth corresponding to the truth established by the second
corollary to the seventeenth proposition of the sixth book, *that
equivalent surfaces upon the same radius will generate equivalent solids*. The synthesis then commences with the conclusion
of the analysis, and retraces its several steps, making that precede which before followed, till we arrive at the required conclusion. Therefore the demonstrations given in the second
corollary to the seventeenth proposition, book sixth, obtain the
conclusion from which the analyses precede. From this it appears that analysis is the instrument of investigation; while
synthesis affords the means of communicating what is already
known; and hence, in the Elements of Euclid, the synthetic
method is followed throughout. What is now said will receive
further illustration from the solution of the following easy
problem.

Given the perimeter and angles of a triangle, to construct it.

Analysis.—Suppose ABC to be the required triangle, and
produce BC both ways, making BD equal to BA, and CE to

CA; then DE is given, for it is equal to the sum of the three sides AB, BC, CA; that is, it is equal to the given perimeter.

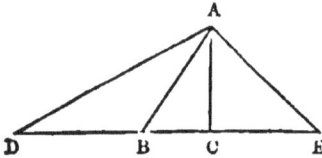

Join AD, AE. Then (I. 1, cor. 1) the angles D and DAB are equal, and therefore each of them is half of ABC, because (I. 20) ABC is equal to both. The angle D therefore is given; and in the same manner it may be shown that E is given, being half of ACB. Hence the triangle ADE is given, because the base DE, and the angles D, E are given; and ADE being given, ABC is also given, the angle DAB being equal to D, and EAC equal to E.

Composition.—Make DE equal to the given perimeter, the angle D equal to the half of one of the given angles, and E equal to the half of another; draw AB, AC, making the angle DAB equal to D, and EAC to E; ABC is the triangle required.

For (I. 1, cor. 2) AB is equal to BD, and AC to CE. To these add BC, and the three, AB, BC, CA, are equal to DE, that is, to the given perimeter. Also (I. 20) the angle ABC is equal to D and DAB, and is therefore double of D, since D and DAB are equal. But D is equal to the half of one of the given angles; therefore ABC is equal to that angle; and, in the same manner, ACB may be proved to be equal to another of the given angles. ABC therefore is the triangle required, since it has its perimeter equal to the given perimeter, and its angles equal to the given angles.

It is impossible to give rules for effecting analyses that will answer in all cases. It may be stated, however, in a general way, that when sums or differences are concerned, the corresponding sums or differences should be exhibited in the assumed figure; that in many cases remarkable points should be joined; or that through them lines may be drawn perpendicular or parallel to remarkable lines, or making given angles with them; and that circles may be described with certain radii, and from certain points as centers; or touching certain lines, or passing through certain points. Some instances of analysis will be given in subsequent propositions; and the student will find it useful to make analyses of many other propositions, such as several in the Exercises.

8. A *porism* is a proposition affirming the possibility of finding such conditions as will render a certain problem indeterminate, or capable of innumerable solutions.

Scho. 3. Porisms may be regarded as having their origin in the solution of problems, which, in particular cases, on account of peculiar relations in their data, admit of innumerable solutions; and the proposition announcing the property or relation which renders the problem indeterminate, is called a porism. This will be illustrated by the solution of the following easy problem.

Through a given point A, let it be required to draw a straight line bisecting a given parallelogram BCDE.

Suppose AFG to be the required line, and let it cut the sides BE, CD in F, G, and the diagonal CE in H. Then from the equivalent figures EBC, FBCG take FBCH, and the remaining triangles EHF, CHG are equal. Now, since (I. 16 and 11) these triangles are equiangular, it is evident that they can be equal in area only when their sides are equal; wherefore H is the middle point of the diagonal. The construction, therefore, is effected by bisecting the diagonal EC in H, and drawing AFHG. For the triangles CHG, EHF are equiangular, and since CH, HE are equal, the triangles are equal. To each of them add the figure FBCH; then the figure FBCG is equivalent to the triangle EBC, that is, to half the parallelogram BD.

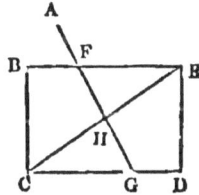

Now, since the diagonal CE is given in magnitude and position, its middle point H is given in position, and therefore H is always a point in the required line, wherever A is taken. Hence, so long as A and H are different points, the straight line AHG (I. post 1) is determined. This, however, is no longer so, if the given point A be the intersection of the diagonals, that is, the point H, as in that case only one point of the required line is known, and the problem becomes indeterminate, any straight line whatever, through H, equally answering the conditions of the problem; and we are thus led by the solution of the problem to the conclusion, that *in a parallelogram a point may be found, such that any straight line whatever drawn through it, bisects the parallelogram ;* and this is a porism.

The seventy-sixth proposition of the Exercises, when considered in a particular manner, affords another instance of a porism ; as it appears that if a circle and a point D or E be given, another point E or D may be found, such that any circle whatever, described through D and E, will bisect the circumference of the given circle; and this may be regarded as the indeterminate case of the problem, in which it is required, through two given points, to describe a circle bisecting the circumference of another given circle,—a problem which is always determinate, except when the points are situated in the manner supposed in the proposition.

9. *Isoperimetrical* figures are such as have their perimeters, or bounding lines, equal.

10. The general problem of the tangencies, as understood by the ancients, is as follows : Of three points, three straight lines, and three circles of given radii, any three being given in position ; it is required to describe a circle passing through the points, and touching the straight lines and circles. This general problem comprehends ten subordinate ones, the data of which are as follows : (1.) three points ; (2.) two points and a straight line ; (3.) two points and a circle ; (4.) a point and two straight lines ; (5.) a point, a straight line, and a circle ; (6.) a point and two circles ; (7.) three straight lines ; (8.) two straight lines and a circle ; (9.) a straight line and two circles ; and (10.) three circles. The first and seventh of these are the second and fifth corollaries of the twenty-fifth proposition of the third book.

If a circle be continually diminished, it may be regarded as becoming ultimately a point. By being continually enlarged, on the contrary, it may have its curvature so much diminished that any portion of its circumference may be made to differ in as small a degree as we please from a straight line. Viewing the subject in this light, we may regard the first nine of the problems now mentioned, as comprehended in the tenth. Thus, we shall have the first, by supposing the circles to become infinitely small ; the seventh, by supposing them infinitely great ;

the fifth, by taking one of them infinitely small, one infinitely great, and one as a circle of finite magnitude; and so on with regard to the others. These views of the subject tend to illustrate it; but they do not assist in the solution of the problems.

Scho. 4. In the fifth problem, the straight line may fall without the circle, may cut it, or may touch it; the point may be without the circle, within it, or in its circumference; or it may be in the given straight line, or on either side of it; and it will be an interesting exercise for the student, in this and many other problems, to consider the variations arising in the solution from such changes in the relations of the data, and to determine what relations make the solution possible, and what render it impossible. It may also be remarked, that in many problems there will be slight variations in the proofs of different solutions of the same problem, even when there is no change in the method of solution; such as in the present instance, when the required circle is touched externally, and when internally. Thus, while in one case angles may coincide, in another the corresponding ones may be vertically opposite; and the reference may sometimes be to the *converse* of the first corollary and sometimes to the *converse* of the second corollary of the eighteenth proposition, book third. It is, in general, unnecessary to point out these variations, as, though they merit the attention of the student, they occasion no difficulty.

PROPOSITIONS.

Prop. I.—Theor.—*If an angle of a triangle be bisected by a straight line, which likewise cuts the base, the rectangle contained by the sides of the triangle is equivalent to the rectangle contained by the segments of the base, together with the square of the straight line bisecting the angle.*

Let ABC be a triangle, and let the angle BAC be bisected by AD; the rectangle BA.AC is equal to the rectangle BD.DC, together with the square of AD.

Describe the circle (III. 25, cor.) ACB about the triangle; produce AD to meet the circumference in E, and join EC. Then (hyp.) the angles BAD, CAE are equal; as are also (III. 18, cor. 1) the angle B and E, for they are in the same segment;

therefore (V. 3, cor.) in the triangles ABD, AEC, as BA : AD :: EA : AC; and consequently (V. 10, cor.) the rectangle BA. AC is equivalent to EA.AD, that is (II. 3), ED.DA, together with the square of AD. But (III. 20) the rectangle ED.DA is

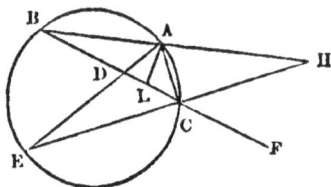

equivalent to the rectangle BD.DC; therefore the rectangle BA.AC is equivalent to BD.DC, together with the square of AD; wherefore, if an angle, etc.

Scho. From this proposition, in connection with the fourth proposition of fifth book, we have the means of computing AD, when the sides are given in numbers. For, by the fourth, BA : AC :: BD : DC; and, by composition, BA+AC : AC :: BC : DC. This analogy gives DC, and BD is then found by taking DC from BC. But by this proposition BA.AC=BD.DC +AD²; therefore from BA.AC take BD.DC, and the square root of the remainder will be AD.

In a similar manner, from the fourth proposition of the fifth book, and the second proposition of these Exercises, the line bisecting the exterior vertical angle may be computed; and from the third proposition of these Exercises, in connection with the eleventh or twelfth of the second book, the diameter of the circumscribed circle may be computed, when the sides of the triangle are given in numbers.

Prop. II.—Theor.—*If an exterior angle of a triangle be bisected by a straight line, which cuts the base produced ; the rectangle contained by the sides of the triangle, and the square of the bisecting line are together equivalent to the rect-*
• *angle contained by the segments of the base intercepted between its extremities and the bisecting line.*

Let, in the foregoing diagram, the exterior angle ACF of the triangle BAC be bisected by HC; the rectangle BC.AC and the square of HC are together equivalent to the rectangle BH.AH.

Describe the circle (III. 25, cor.) ABEC about the triangle BAC; produce HC (I. post. 2) to E, and join EA. Then, since (hyp.) the angles FCH and HCA are equal, their supplements FCE and ACE (I. def. 20 and ax. 3) are also equal; and (III. 18, cor. 1) B and E are equal. Therefore (V. 3, cor.) in the triangles BCH and EAC, BC : CH :: EC : AC, and consequently (V. 10, cor.) the rectangle BC.AC is equivalent to EC.CH. To each add square of CH; therefore BC.AC+CH² are equivalent to EC.CH+CH²; or (II. 3) BC.AC+CH² are equivalent to EH.CH; or (III. 21, cor.) BC.AC+CH² are equivalent to BH.AH. Therefore, if an exterior angle, etc.

PROP. III.—THEOR.—*If from an angle of a triangle a perpendicular be drawn to the base; the rectangle contained by the sides of the triangle is equivalent to the rectangle contained by the perpendicular and the diameter of the circle described about the triangle.*

Also, in the foregoing diagram, let ABC be a triangle, AL the perpendicular from the angle A to BC; and AE a diameter of the circumscribed circle ABEC; the rectangle BA.AC is equivalent to the rectangle AL.AE.

Join EC. Then the right angle BLA is equal (III. 11) to the angle ECA in a semicircle, and (III. 18, cor. 1) the angle B to the angle E in the same segment; therefore (V. 3, cor.) as BA : AL :: EA : AC; and consequently (V. 10, cor.) the rectangle BA.AC is equivalent to the rectangle EA.AL. If, therefore, from an angle of a triangle, etc.

PROP. IV.—THEOR.—*The rectangle contained by the diagonals of a quadrilateral inscribed in a circle, is equivalent to both the rectangles contained by its opposite sides.*

Let ABCD be a quadrilateral inscribed in a circle, and join AC, BD; the rectangle AC.BD is equivalent to the two rectangles AB.CD and AD.BC.

Make the angle ABE equal to DBC, and take each of them from the whole angle ABC; then the remaining angles CBE, ABD are equal; and (III. 18, cor. 1) the angles ECB, ADB are equal. Therefore (V. 3, cor.) in the triangles ABD, EBC,

as BC : CE :: BD : DA; whence (V. 10, cor.) BC.DA=
BD.CE. Again: in the triangles BAE, BDC, because (const.)
the angles ABE, DBC are equal, as also
(III. 18, cor. 1) BAE, BDC; therefore
(V. 3, cor.) as BA : AE :: BD : DC;
whence (V. 10, cor.) BA.DC=BD.AE.
Add these equivalent rectangles to the
equivalents BC.DA and BD.CE; then
BA.DC+BC.DA=BD.CE+BD.AE, or
(II. 1) BA.DC+BC.DA=BD.AC. There-
fore, the rectangle, etc.

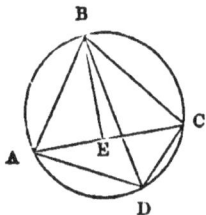

Cor. 1. If the sides AD, DC, and consequently (III. 16, cor.
1) the arcs AD, DC, and the angles ABD, CBD be equal, the
rectangle BD.AC is equivalent to AB.AD together with BC.AD,
or (II. 1) to the rectangle under AD, and the sum of AB and BC.
Hence (V. 10, cor.) AB+BC : BD :: AC : AD or DC.

Cor. 2. If AC, AD, CD be all equal, the last analogy becomes AB+BC : BD :: AD : AD; whence AB+BC=BD.
Hence in an equilateral triangle inscribed in a circle, a straight
line drawn from the vertex to a point in the arc cut off by the
base is equal to the sum of the chords drawn from that point
to the extremities of the base.

PROP. V.—THEOR.—*The diagonals of a quadrilateral inscribed in a circle, are proportional to the sums of the rectangles contained by the sides meeting at their extremities.*

Let ABCD be a quadrilateral inscribed in a circle, and AC,
BD its diagonals; AC : BD :: BA.AD+
BC.CD : AB.BC+AD.DC.

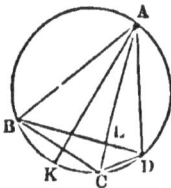

If AC, BD cut one another perpendicularly in L, then (Ex. 2) AK being the diameter of the circle, AL.AK=BA.AD, and CL.
AK=BC.CD; whence, by addition (I. ax.
2), AL.AK+CL.AK, or (II. 1) AC.AK=
BA.AD+BC.CD. In a similar manner,
it would be proved that BD.AK=AB.BC
+AD.DC. Hence AC.AK : BD.AK, or (V. 1) AC : BD ::
BA.AD+BC.CD : AB.BC+AD.DC.

But if AC be not perpendicular to BD, draw AEF perpen-

dicular and CF parallel to BD, and DGH perpendicular and
BH parallel to AC. Then, because EF is equal to the perpen-
dicular drawn from C to BD, and GH equal to the one drawn
from B to AC; it would be prov-
ed as before, that AF.AK=BA.
AD+BC.CD, and DH.AK=
AB.BC+AD.DC. Hence, AF.
AK : DH.AK, or (V. 1) AF :
DH :: BA.AD+BC.CD : AB.
BC+AD.DC. But the triangles
AFC, DHB are equiangular, hav-
ing the right angles F and H, and
the angles ACF, DBH, each equal
(I. 16) to ALD; therefore (V. 3)
AF : AC :: DH : DB, and alternately AF : DH :: AC :
DB. Hence (IV. 7) the foregoing analogy becomes AC : BD
:: BA.AD+BC.CD : AB.BC+AD.DC. Wherefore, the
diagonals, etc.

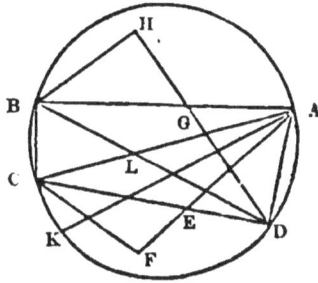

Scho. From this proposition and the last, when the sides of
a quadrilateral inscribed in a circle are given, we can find the
ratio of the diagonals and their rectangle, and thence (V. 15)
the diagonals themselves. Also, if the sides be given in numbers, we can compute the diagonals. Thus, let the sides taken
in succession round the figure be 50, 78, 104, and 120. Then,
the ratio of the diagonals will be that of $50 \times 78 + 104 \times 120$ to
$50 \times 120 + 78 \times 104$; that is, 16380 to 14112, or 65 to 56, by
dividing by 252. Again, the rectangle of the diagonals is $50 \times
104 + 78 \times 120$, or 14560. But similar rectilineal figures are as
the squares of the corresponding sides, and consequently the
sides are as the square roots of the areas; therefore, taking 65
and 56 as the sides of a rectangle, we have its area equal to
3640; and $\sqrt{3640}$ is to $\sqrt{14560}$, or $\sqrt{3640}$ is to $\sqrt{(4 \times 3640)}$,
that is, 1 : 2 :: 65 : 130 :: 56 : 112; so that 130 and 112
are the diagonals.

PROP. VI.—THEOR.—*If in a straight line drawn through
the center of a circle, and on the same side of the center, two
points be taken so that the radius is a mean proportional between their distances from the center ; two straight lines drawn*

15

from those points to any point whatever in the circumference, are proportional to the segments into which the circumference divides the straight line intercepted between the same points.

Let ABC be a circle, and CAE a straight line drawn through its center D; if ED : DA :: DA : DF, and if BE, BF be drawn from any point B of the circumference; EB : BF :: EA : AF.

Join AB, BD. Then, since DB is equal to DA, we have

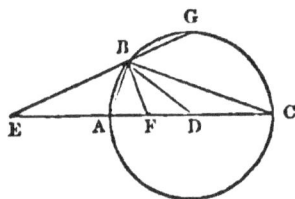

(hyp.) ED : DB :: DB : DF. The two triangles EDB, BDF, therefore, have their sides about the common angle D proportional; wherefore (V. 6) the angle E is equal to FBD. Now the angle BAD is equal (I. 20) to the two angles E and EBA, and also (I. 1, cor.) to ABD; wherefore E and EBA are (I. ax. 1) equal to ABD. From these take the equal angles E and FBD, and (I. ax. 3) the remaining angles EBA, ABF are equal; and therefore (V. 4) in the triangle EBF, EB : BF :: EA : AF. If, therefore, in a straight line, etc.

Cor. Join BC, and produce EB to G. Then, since ABC is (III. 11) a right angle, it is equal to the two EBA, CBG. From these equals take the equal angles ABF, EBA, and the remainders FBC, CBG are equal; and therefore (V. 4, 2d case) EB : BF :: EC : CF. But it has been proved that EB : BF :: EA : AF; therefore (IV. 7) EA : AF :: EC : CF. Hence, if the segments EA, AF be given, the point C may be determined by the method shown in the third corollary to the sixteenth proposition of the first book; and the circle ABC may then be described, its diameter AC being determined.

Scho. The circle may also be determined in the following manner: Since (hyp.) ED : DA :: DA : DF, by division, EA : DA :: AF : DF; whence, alternately and by division, EA—AF : AF :: AF : DF. Hence DF is a third proportional to the difference of EA, AF, and to AF, the less; and thus the center D is determined. From the last analogy also we obtained (IV. 11) EA—AF : EA :: AF : AD; an analogy which serves the same purpose, since it shows that the radius of the circle is a fourth proportional to the difference of the segments EA, AF, and to those segments themselves.

PROB. VII.—THEOR.—*The perpendiculars drawn from the three angles of any triangle to the opposite sides, intersect one another in the same point.*

If the triangle be right-angled, it is plain that all the perpendiculars pass through the right angle. But if it be not right-angled, let ABC be the triangle, and about it describe a circle; then, B and C being acute angles, draw ADE perpendicular to BC, cutting BC in D, and the circumference in E; and make DF equal to DE; join BF and produce it, if necessary, to cut AC, or AC produced in G; BG is perpendicular to AC. Join BE; and because FD is equal to DE, the angles at D right angles, and DB common to the two triangles FDB, EDB, the angle FBD is equal (I. 3) to EBD; but (III. 18, cor. 1) CAD, EBD are also equal, because they are in the same segment; therefore CAD is equal to FBD or GBC. But the angle ACB is common to the two triangles ACD, BCG; and therefore (I. 20, cor. 5) the remaining angles ADC, BGC are equal; but (const.) ADC is a right angle; therefore also BGC is a right angle, and BG is perpendicular to AC. In the same manner it would be shown that a straight line CH, drawn through C and F, is perpendicular to AB. The three perpendiculars therefore all pass through F; wherefore, the perpendiculars, etc.

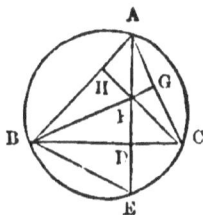

Scho. This limitation prevents the necessity of a different case, which would arise if the perpendicular AD fell without the triangle. If the angle A be obtuse, the point F lies without the circle, and BF, not produced, cuts AC produced. The proof, however, is the same, and it is very easy and obvious. Another easy and elegant proof, of which the following is an outline, is given in Garnier's "Reciproques," etc., Theor. III., page 78: Draw BG and CH perpendicular to AC and AB; join GH, and about the quadrilaterals AHFG and BHGC describe circles, which can be done, as is easily shown; draw also AFD. Then the angles BAD, BCH are equal, each of

them being equal (III. 18, cor. 1) to HGF ; and the angle ABC
being common, ADB is equal (I. 20) to BHC, and is therefore
a right angle.

PROP. VIII.—THEOR.—*From* AB, *the greater side of the triangle* ABC, *cut off* AD *equal to* AC, *and join* DC; *draw* AE
bisecting the vertical angle BAC, *and join* DE; *d ino also* AF
perpendicular to BC, *and* DG *parallel to* AE. *Then* (1.) *the
angle* DEB *is equivalent to the difference of the angles at the
base,* ACB, ABC ; *or of* BAF, CAF, *or of* AEB, AEC ; *and*
DE *is equal to* EC ; (2.) *the angles* BCD, EAF *are each equivalent to half the same difference ;* (3.) ADC *or* ACD *is equivalent to half the sum of the angles at the base, or to the complement of half the vertical angle ;* (4.) BG *is equivalent to the
difference of the segments* BE, EC, *made by the line bisecting
the vertical angle.*

1. In the triangles AED, AEC, AD, AC are equal, AE common, and the contained angles equal ; therefore (I. 3) DE is
equal to EC, the angle ADE to ACE, and AED to AEC. But
(I. 20) because BD is produced, the
angle ADE is equivalent to B and
BED ; therefore BED is the difference of B and ADE, or of B and
ACB. Also BED is the difference
of AEB, AED, or of AEB, AEC.
Again : ABF, BAF are equivalent
to ACF, CAF, each pair being (L
20, cor. 3) equivalent to a right angle. Take away ABF ;
then, because the difference of ACF, ABF is BED, there remains BAF equivalent to BED, CAF ; that is, BED is the
difference of BAF, CAF.

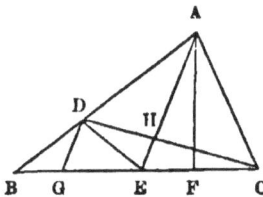

2. The difference BED is equivalent (I. 20) to the two angles
ECD, EDC, which (I. 1, cor.) are equal ; therefore ECD is half
of BED. Again : in the triangles AHD, AHC, because AD,
AC are equal, AH common, and the contained angles equal,
DH is equal (I. 3) to HC, and the angles at H are equal, and
are therefore right angles. Then, in the triangles AEF, CEH,
the angles AFE, CHE are equal, and AEC common ; therefore

(I. 20, cor. 5) the angle EAF is equal to ECII, which has been proved to be equivalent to the half of BED.

3. Since the angle BAC is common to the triangles ABC, ADC, the angles ADC, ACD are (I. 20) equal to ABC, ACB; and therefore, since ADC, ACD are equal, each of them is equivalent to half the sum of ABC, ACB; also either of them, ADC, is the complement of DAII, half the vertical angle, since AIID is a right angle.

4. Because DII, IIC are equal, and IIE, DG parallel, GE is equal (V. 2) to EC; and therefore BG, the difference of BE, GE, is also the difference of BE, EC.

Scho. 1. It is easy to prove without proportion, that if AB (in the figure for the above proposition) be bisected in D, the straight line DE parallel to BC bisects AC, and that the triangle ADE is a fourth of ABC. For (I. 15, cor. 5) the triangles BDC, BEC are equal. But (I. 15, cor. 5) BDC is half of ABC; and therefore BEC is half of ABC, and is equal to BEA. Hence the bases AE, EC are equal; for if they were not, the triangles ABE, CBE (I. 15, cor. 6) would be unequal. Again: because AD, DB are equal, the triangle ADE is (I. 15, cor. 5) half of ABE, and therefore a fourth of ABC.

Conversely, if DE bisect AB, AC, it is parallel to BC. For (I. 15, cor.) the triangles BDC, BEC are each half of ABC; and these being therefore equal, DE is parallel (I. 15, cor. 5) to BC.

Hence, it is plain (I. 14) that the straight lines joining DE, DF, EF divide the triangle ABC into four equal triangles, similar to the whole and to one another; and that each of these lines is equal to half the side to which it is parallel.

Scho. 2. Instead of cutting off AD equal to AC, AC may be produced through C, and by cutting off, on AC thus produced, a part terminated at A, and equal to AB, and by making a construction similar to that of the foregoing proposition, it will be easy to establish the same properties as those above demonstrated, or ones exactly analogous.

PROP. IX.—PROB.—*Given the base of a triangle, the differ-ence of the sides, and the difference of the angles at the base; to construct it.*

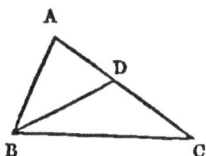

Make BC equal to the given base, and CBD equal to half the difference of the angles at the base; from C as center, at a distance equal to the difference of the sides, describe an arc cutting BD in D; join CD and produce it; make the angle DBA equal to BDA; ABC is the required triangle.

For (I. 1, cor.) AD is equal to AB, and the difference of AC, AD, or of AC, AB, is CD; and (Ex. 8) since AD is equal to AB, CBD is equal to half the difference of the angles at the base. The triangle ABC, therefore, is the one required, as it has its base equal to the given base, the difference of its sides equal to the given difference, and the difference of the angles at the base equal to the given difference of those angles.

Method of Computation. In the triangle BCD there are given BC, CD, and the angle CBD; whence the angle C can be computed; and the sum of this and twice CBD is the angle ABC. Then, in the whole triangle ABC, the angles and the side BC are given; whence the other sides may be computed; or, one of them being computed, the other will be found by means of the given difference CD.

PROP. X.—PROB.—*Given the segments into which the base of a triangle is divided by the line bisecting the vertical angle and the differeace of the sides ; to construct the triangle.*

Construct the triangle CED, having the sides CE, ED equal to the given segments, and CD equal to the given difference of the sides; produce CE, and make EB equal to ED; bisect the angle BED by EA, meeting CD produced in A, and join AB; ABC is the required triangle.

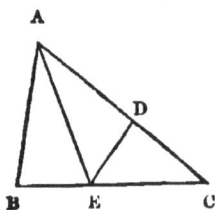

For, in the triangles AEB, AED, BE is equal to ED, EA common, and the angle BEA equal to DEA; therefore (I. 3) BA is equal to DA, and the angle EAB to EAD. Hence, ABC is the required triangle; for CD, the difference of its sides, is equal to the given difference, and BE, EC, the segments into which the base is divided by the line bisecting the vertical angle, are equal to the given segments.

Meth d of Computation. The sides of the triangle CDE are given, and therefore its angles may be computed; one of which and the supplement of the other are the angles C and B. Then, in the triangle ABC, the angles and BC are given, to compute the remaining sides.

Since (V. 4) CE : EB :: CA : AB, we have, by division, CE—EB : EB :: CA—AB : AB ; which, therefore, becomes known, since the first three terms of the analogy are given; and thence AC will be found by adding to AB the given difference of the sides.

Prop. XI.—Prob.—*Given the base of a triangle, the vertical angle, and the difference of the sides ; to construct the triangle.*

Let MNO be the given vertical angle; produce ON to P, and bisect the angle MNP by NQ. Then, make BD equal to the difference of the sides, and the angle ADC equal to QNP; from B as center, with the given base as radius, describe an arc cutting DC in C; and make the angle DCA equal to ADC; ABC is the required triangle. For (const.) the angles ACD, ADC are equal to MNP, and therefore (I. 20 and 9) the angle A is equal to MNO. But (I. 1, cor.) AD is equal to AC, and therefore BD is the difference of the sides AB, AC; and the base BC is equal to the given base ; wherefore ABC is the triangle required.

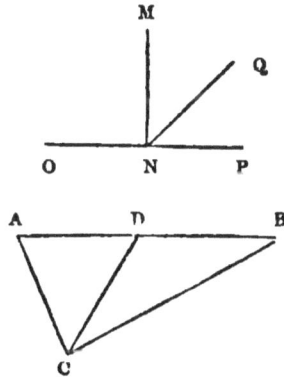

Method of Computation. In the triangle CBD, the sides BC, BD, and the angle BDC, the supplement of ADC or MNQ are given; whence the other angles can be computed. The rest of the operation will proceed as in the ninth proposition of these Exercises.

Prop. XII.—Prob.—*Given one of the angles at the base of a triangle, the difference of the sides, and the difference of the segments into which the base is divided by the line bisecting the vertical angle; to construct the triangle.*

Construct the triangle DBG, having the angle B equal to the given angle, BD equal to the difference of the sides, and BG equal to the difference of the segments; draw DC perpendicular to DG, and meeting BG produced in C; produce BD, and make the angle DCA equal to CDA; ABC is the triangle required. For it has B equal to the given angle, and the difference of its sides BD equal to the given difference; and if AHE be drawn bisecting the angle BAC, it bisects (I. 3) CD, and is perpendicular to it; it is therefore parallel to DG, one side of the triangle CDG; and, bisecting CD in H, it also (V. 2) bisects GC in E. Hence BG, the difference of BE, GE is also the difference of BE, EC, the segments into which the base is divided by the line bisecting the vertical angle.

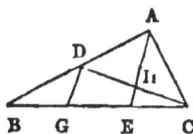

Method of Computation. In the triangle DBG, BD, BG, and the angle B are given; whence (Trig. 3) we find half the difference of the angles BGD, BDG, which is equal to half the angle C. Then (by the same proposition) we have in the triangle ABC, tan $\frac{1}{2}$ (C−B) : tan $\frac{1}{2}$ (C+B) :: c−b or BD : c+b; whence the sides c and b become known, and thence BC by the first case.

For (I. 20) BEA = $\frac{1}{2}$ A+C, and consequently BEA−$\frac{1}{2}$ A = C. But (I. 16) BGD = BEA, and BDG = BAE = $\frac{1}{2}$ A ; and therefore BGD − BDG = C.

Prop. XIII.—Prob.—*Given the base of a triangle, the vertical angle, and the sum of the sides; to construct it.*

Make BD equal to the sum of the sides, and the angle D equal to half the vertical angle; from B as center, with the base as radius, describe an arc meeting DC in C; and make the angle DCA equal to D; ABC is the required triangle.

For (I. 1, cor.) AD is equal to AC, and therefore BA, AC are equal to BD, the given sum. Also (I. 20) the exterior angle BAC is equal to the two D and ACD, or to the double of D, because D and ACD are equal; therefore, since D is half the given vertical angle, BAC is equal to that angle. The triangle ABC, therefore, has its base equal to the given base, its vertical angle equal to the given one, and the sum of its sides equal to the given sum; it is therefore the triangle required.

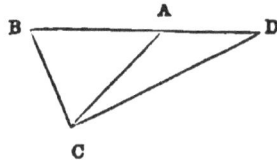

Scho. Should the circle neither cut nor touch DC, the problem would be impossible with the proposed data. If the circle meet DC in two points, there will be two triangles, each of which will answer the conditions of the problem. These triangles, however, will differ only in position, as they will be on equal bases, and will have their remaining sides equal, each to each. This problem might also be solved by describing (III. 19) on the given base BC a segment of a circle containing an angle equal to half the vertical angle; by inscribing a chord BD equal to the sum of the sides; by joining DC; and then proceeding as before. The construction given above is preferable.

Method of Computation. In the triangle BDC, the angle D, and the sides BC, BD are given; whence the remaining angles can be computed; and then, in the triangle ABC, the angles and the side BC are given, to compute the other sides.

PROP. XIV.—PROB.—*Given the vertical angle of a triangle, and the segments into which the line bisecting it divides the base ; to construct it.*

In the straight line BC, take BH and CH equal to the segments of the base; on BC describe (III. 19) the segment BAC containing an angle equal to the vertical angle, and complete the circle; bisect the arc BEC in E; draw EHA, and join BA, CA; ABC is the required triangle. For (III. 16, cor. 1) the angles BAH, CAH are equal, because the arcs BE, EC are equal; and therefore the triangle ABC manifestly answers the conditions of the question.

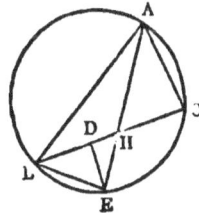

Scho. The construction might also be effected by describing on BH and CH segments each containing an angle equal to half the vertical angle, and joining their point of intersection A with B and C. Another solution may be obtained by the principle (V. 4) that BA · AC :: BH : HC. For if a triangle be constructed having its vertical angle equal to the given one, and the sides containing it equal to the given segments, or having the same ratio, that triangle will be similar to the required one; and therefore on CB construct a triangle equiangular to the one so obtained.

Method of Computation. Join BE, and draw ED perpendicular to BC. Then BD or DC is half the sum of the segments BH, HC, and DH half their difference; and BD is to DH, or twice BD to twice DH, as tan DEB to tan DEH. Now it is easy to show that BED is half the sum of the angles ABC, ACB, and HED half their difference; and therefore these angles become known; and BC being given, the triangle ABC is then resolved by the method for the first case.

Cor. Hence, we have the method of solving the problem in which *the base, the vertical angle, and the ratio of the sides of a triangle are given, to construct it.* For (V. 4) the sides being proportional to the segments BH, HC, it is only necessary to divide the given base into segments proportional to the sides and then to proceed as above. '

PROP. XV.—PROB.—*Given the base, the perpendicular, and the vertical angle of a triangle ; to construct it.*

Make BC equal to the given base, and (III. 10) on it describe a segment capable of containing an angle equal to the vertical angle; draw AK parallel to BC, at a distance from it equal to the given perpendicular and meeting the arc in A; join AB, AC; ABC is evidently the triangle required.

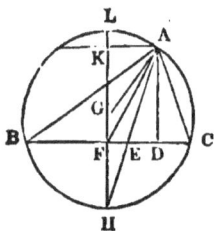

Method of Computation. Draw the perpendicular AD, and parallel to it draw LGH, through the center G; join BG, AG, AH. Now, since AH evidently bisects the angle BAC, the angle HAD or

H is (Ex. 8) equal to half the difference of the angles ABC, ACB, and therefore (III. 10) AGK is the whole difference of those angles. Then, in the right-angled triangle BFG, the angles and BF being known, FG can be computed; from which and from AD or KF, KG becomes known. Now, to the radius BG or AG, FG is the cosine of BGF, or BAC, and KG the cosine of AKG; and therefore FG : KG :: cos BAC : cos AGK. Hence the angles ABC, ACB become known, and thence the remaining sides.

Scho. Should the parallel AK not meet the circle, the solution would be impossible, as no triangle could be constructed having its base, perpendicular, and vertical angle of the given magnitudes. If the parallel cut the circle, there will be two triangles, either of which will answer the condition of the question. They will differ, however, only in position, as their sides will be equal, each to each. If the parallel touch the circle, there will be only one triangle; and it will be isosceles.

PROP. XVI.—PROB.— *Given the base of a triangle, the vertical angle, and the radius of the inscribed circle; to construct the triangle.*

Let GHK be the given angle; produce GH to L, and bisect LHK by HM; on the given base BC describe (III. 19) the segment BDC, containing an angle equal to GHM; draw a straight line parallel to BC, at a distance equal to the given radius, and meeting the arc of the segment in D; join DB, DC; and make the angles DBA, DCA equal to DBC, DCB, each to each; ABC is the required triangle.

Produce BD to E, and draw DF perpendicular to BC. Then, since (const.) the angle BDC is equal to GHM, the two DBC, DCB are equal (I. 20 and 9) to LHM, and therefore (const.) ABD, ACD are equal to KHM. But (I. 20) BDC is equal to BEC, ECD, or to BAC, ABD, ACD, because (I. 20) BEC is equal to BAC, ABD. Therefore BAC, ABD, ACD are equal to GHM; from the former take ABD, ACD, and from the latter KHM, which is equal to

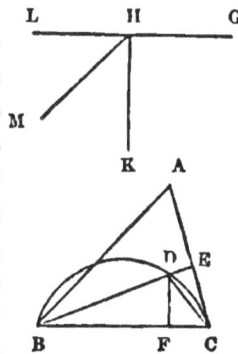

them, and the remainders BAC, GHK are equal. It is plain,
also (I. 14), that perpendiculars drawn from D to AB and AC
would be each equal to DF; and therefore a circle described
from D as center, with DF as radius, would be inscribed in the
triangle ABC; and BC being the given base, and A being
equal to the given vertical angle, ABC is the required triangle.

The method of computation is easily derived from that of the
preceding proposition.

Scho. Should the parallel to BC not meet the arc of the segment, the solution would be impossible, as there would be no
triangle which could have its base, its vertical angle, and its
inscribed circle of the given magnitudes. If the parallel be a
tangent to the arc, the radius of the inscribed circle would be
a maximum. Hence, to solve the problem in which the base
and the vertical angle are given, to construct the triangle, so
that the inscribed circle may be a maximum, describe the segment as before, and to find D bisect the arc of the segment.
The rest of the construction is the same as before; and the triangle will evidently be isosceles.

PROP. XVII.—PROB.—*Given the three lines drawn from the
vertex of a triangle, one of them perpendicular to the base, one
bisecting the base, and one bisecting the vertical angle ; to construct the triangle.*

Take any straight line BC and draw DA perpendicular to it,
and equal to the given perpendicular; from A as center, with
radii equal to the lines bisecting the vertical angle and the
base, describe arcs cutting BC in E and
F, and draw AEH and AF; through F
draw GFH perpendicular to BC, and
draw AG making the angle HAG equal
to H, and cutting HG in G; from G as
center, with GA as radius, describe a
circle cutting BC in B and C; join AB,
AC; ABC is the triangle required.

For (III. 2) since GFH is perpendicu-
lar to BC, BC is bisected in F; and (III. 17) the arcs BH, HC
are equal. Therefore (III. 16, cor. 1) the angles BAH, CAH

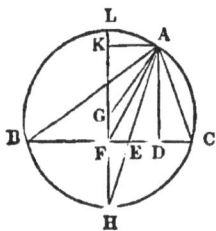

arc equal. Hence, in the triangle ABC, the perpendicular AD, the line AE bisecting the vertical angle, and the line AF bisecting the base, are equal to the given lines. Therefore ABC is the triangle required.

Scho. If the three given lines be equal, the problem is indeterminate; as any isosceles triangle, having its altitude equal to one of the given lines, will answer the conditions.

Method of Computation. Through A draw a parallel to BC, meeting HG produced in K. Then, in the right-angled triangle ADE, AE, AD being given, DAE, or H, may be computed; the double of which is AGK; and AK or FD being given, AG, GK can be found, and thence GF. Hence, if GB were drawn, it and GF being known, BF, and the angle BGF, or BAC, can be computed. The rest is easy; DAE, half the difference of B and C, being known.

Analysis. Let ABC be the required triangle, AD the perpendicular, and AE, AF the lines bisecting the vertical angle and the base. About ABC describe (III. 25, cor. 2) a circle, and join its center G, with A and F, and produce GF to meet the circumference in H. Then (III. 2) GFH is perpendicular to BC, and (III. 17) the arcs BH, HC are equal. But (III. 16) the equal angles BAE, CAE at the circumference stand on equal arcs; and therefore AE being produced, will also pass through H; and the point H, and the angle GHA and its equal HAG are given. Hence also the center G and the circle are given, and the method of solution is plain.

Prop. XVIII.—Prob.—*Given the base of a triangle, the vertical angle, and the straight line bisecting that angle ; to construct the triangle.*

On the given base BC describe (III. 19) the segment BAC capable of containing an angle equal to the given vertical angle, and complete the circle; bisect the arc BEC in E, and join EC; perpendicular to this, draw CF equal to half the line bisecting the vertical angle, and from F as center, with FC as radius, describe the circle CGH, cutting the straight line passin through E and F in G and H; make ED equal to EG, and draw EDA; lastly, join AB, AC, and ABC is the required triangle.

For the triangles CEA, CED are equiangular, the angle CEA being common, and BCE, CAE being each equal to BAE. Therefore AE : EC :: EC : ED, and (V. 9, cor.) AE.ED=

EC². But (III. 16, cor. 3, and 21) HE.EG or HE.ED=EC²; and therefore AE.ED=HE.ED; whence AE=HE, and (I. ax. 3) AD=GH=2CF. AD is therefore equal to the given bisecting line, and it bisects the angle BAC. Hence ABC is the required triangle.

Method of Computation. Draw EL perpendicular to BC, and join CH. Then BCE is equal to BAE, half the vertical angle A; and therefore, to the radius EC; CL is the cosine of $\frac{1}{2}$ A, and CF is the tangent of CEF to the same radius; wherefore, to any radius, CL : CF, or BC : AD :: cos $\frac{1}{2}$ A : tan CEF or cot EFC; and hence the angle H, being half of EFC, is known. Also ECH is the complement of H, because ECF is a right angle, and FCH equal to H. But (TRIG. 2) EC : EH or EA :: sin H : sin ECH, or cos H; or (TRIG. defs. cor. 6) EC : EA :: R : cot H. Also in the triangle ACE, EC : EA :: sin $\frac{1}{2}$ A : sin ACE; whence (IV. 7) R : cot H :: sin $\frac{1}{2}$ A : sin ACE; whence ACE may be found; and if from it, and from ABE, its supplement (BE being supposed to be joined) BCE be taken, the remainders are the angles at the base.

Analysis. Let ABC be the required triangle, and let AD, the line bisecting the vertical angle, be produced to meet the circumference of the circumscribed circle in E; join also EC. Then (III. 19) the circumscribed circle is given, since the base and vertical angle are given; and the arc BEC is given, as are also its half EC, and the chord EC. Now the triangles AEC,

DEC are equiangular; for the angle CEA is common, and (III. 18, cor. 1) BCE is equal to BAE or EAC. Hence AE : EC :: EC : ED, and therefore $AE.ED = EC^2$. Hence (III. 21) it is evident that if EC be made a tangent to a circle, and if through the extremity of the tangent a line be drawn cutting the circle, so that the part within the circle may be equal to AD, DE will be equal to the external part; whence the construction is manifest.

PROP. XIX.—PROB.—*Given the straight lines drawn from the three angles of a triangle to the points of bisection of the opposite sides; to construct the triangle.*

Trisect the three given lines, and describe the triangle ABC having its three sides respectively equal to two thirds of the three given lines; complete the parallelogram ABEC, and draw the diagonal AE; produce also CB, making BF equal to BC; and join FA, FE; AFE is the required triangle.

Produce AB, EB to G, H. Then (II. 14, cor.) AE, BC bisect each other in D, and therefore FD is equal to one of the given lines, for BD is one third of it, and FB two thirds. Again: because FB, BC are equal, and HB parallel to AC, FA is bisected in H, and HB is half of AC or BE. Hence, HE is equal to another of the given lines, and it bisects FA. In the same manner it would be proved, that AG is equal to the remaining line, and that it bisects FE. Hence FAE is the triangle required.

Method of Computation. BD, which is a third of one of the given lines, bisects AE, a side of the triangle ABE, in which the sides AB, BE are respectively two thirds of the two remaining lines. Then (I. 24, cor. 6, and II. 12, cor.) $2AD^2 = AB^2 + BE^2 - 2BD^2$; whence AD, and consequently AE may be found; and in the same manner the other sides may be computed.

PROP. XX.—PROB.—*Given the three perpendiculars of a triangle; to construct it.*

Let A, B, C be three given straight lines; it is required to describe a triangle having its three perpendiculars respectively equal to A, B, C.

Take any straight line D, and describe a triangle EFG. having the sides FG, FE, EG third proportionals to A and D, B and D, C and D; and draw the perpendiculars EH, GL, FK.

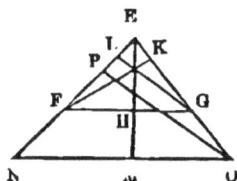

Then the rectangles FG.A, EF.B are equal, each being equal to the square of D; and therefore EF : FG :: A : B. But in the similar triangles EHF, GLF, EF : FG :: EH : GL; wherefore EH : GL :: A : B; and in the same manner it would be proved, that EH : FK :: A : C. Hence (IV. 7) if EH be equal to A, GL is equal to B, and FK to C; and EFG is the triangle required.

But if EH be not equal to A, make EM equal to it, and draw NMO parallel to FG, and meeting EF, EG, produced, if necessary, in N and O; ENO is the required triangle. Draw OP perpendicular to EN. Then EM : EH :: EO : EG, and OP : GL :: EO : EG; whence (IV. 7 and alternately) EM : OP :: EH : GL; or, by the foregoing part, EM : OP :: A : B; wherefore (IV. 7) since EM is equal to A, OP is equal to B; and it would be proved in a similar manner, that the perpendicular from N to EO is equal to C.

Method of Computation. By dividing any assumed number successively by A, B, C, we find the sides of the triangle EFG, and thence its angles, or those of ENO; whence, since the perpendicular EM is given, the sides are easily found.

Or, when the sides of EFG are found, its perpendicular EH may be computed in the manner pointed out in the note to the twelfth proposition of the second book. Then EH : A :: FG : NO :: EF : EN :: EG : EO.

PROP. XXI.—PROB.—*Given the sum of the legs of a right-angled triangle, and the sum of the hypothenuse, and the perpendicular to it from the right angle; to construct the triangle.*

Let the sum of the legs of a right-angled triangle be equal to the straight line A, and the sum of the hypothenuse and perpendicular equal to BC; it is required to construct the triangle.

Find (I. 24, cor. 4) a straight line the square of which is equal to the excess of the square of BC above that of A, and cut off BD equal to that line; on DC as diameter describe a semicircle, and draw EF parallel to DC at a distance equal to BD; join either point of intersection, E, with D and C; DEC is the required triangle.

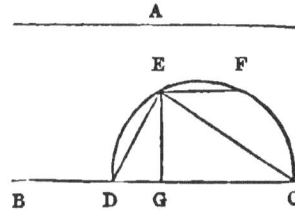

Draw the perpendicular EG, which (const.) is equal to BD. Then (II. 4) BC^2 or $A^2 + EG^2 = (DC + EG)^2 = DC^2 + EG^2 + 2DC.EG$; whence $A^2 = DC^2 + 2DC.EG$. Also $(DE + EC)^2 = DE^2 + EC^2 + 2DE.EC = DC^2 + 2DC.EG$, because (III. 11, and I. 24, cor. 1) $DC^2 = DE^2 + EC^2$, and $DC.GE = DE.EC$, each being equal to twice the area of the triangle DEC. Hence $(DE + EC)^2 = A^2$; wherefore (I. 23, cor. 3) $DE + EC = A$; and DEC is the triangle required.

Method of Computation. From the construction, we have BD or $EG = \sqrt{(BC^2 - A^2)}$. Then $DC = BC - BD$; by halving which we get the radius of the semicircle; and if from the square of the radius drawn from E, the square of EG be taken, and if the square root of the remainder be successively taken from the radius, and added to it, the results will be the segments DG, GC; from which, and from EG, the sides (I. 24, cor. 1) are readily computed.

PROP. XXII.—PROB.— *Given the base of a triangle, the perpendicular, and the difference of the sides ; to construct it.*

Make AB equal to the given base, and parallel to it draw CD, at a distance equal to the given perpendicular; draw BDF perpendicular to CD, and make DF equal to DB; from A as center, with a radius AE equal to the given difference, describe the circle ELN; through B, F describe any circle cutting ELN in L, N, and let G be the point in which a straight line drawn through L, N cuts FB produced; draw the tangent GK, and

16

draw AKM cutting CD in M; join BM; and it is evident from the second corollary to the ninth proposition of the third book, that AMB is the required triangle.

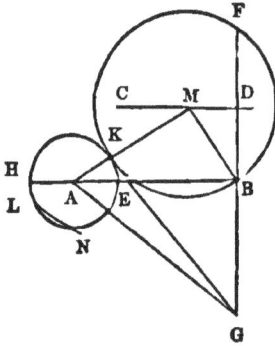

Method of Computation. From M as center, with MK as radius, describe a circle, and by the corollary referred to, it will pass through B and F. Join AG and produce BA to H. Then the rectangle FG.GB=AG^2—AK^2, each being equal to the square of GK ; that is,

$$FB.BG+BG^2=AB^2+BG^2-AK^2.$$

Take away BG^2; then FB.BG=AB^2—AK^2=(AB+AK) (AB —AK)=HB.EB. Hence BG becomes known. Then, in the two right-angled triangles GBA, GKA, the angles at A can be computed, and their difference is the angle MAB in the required triangle. The rest is easy, if the perpendicular from M to AB be drawn.

Prop. XXIII.—Prob.—*Given the base, the area, and the ratio of the sides of a triangle ; to construct it.*

Let AB be the given base, and (V. 3, and scho.) find the points C, D, so that AC, CB, and AD, DB are in the ratio of the sides ; on CD as diameter describe the semicircle CED, and (II. 5, scho.) to AB apply a parallelogram BF double of the given area ; let FG, the side of this opposite to AB, produced if necessary, cut the semicircle in E, and join EA, EB; EAB is the required triangle.

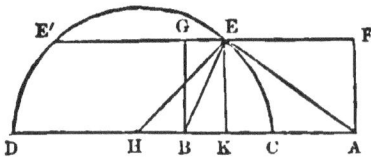

For (Ex. 6, cor.) AE is to EB as AC to CB; that is, in the given ratio. Also (I. 15, cor. 4) AEB is half of the parallelogram BF, which is double of the given area. Therefore AEB is on the given base, is of the given area, and has its sides in the given ratio.

Scho. If FG, or FG produced, do not meet the semicircle,

the problem is impossible; if it cut it in E and E', there will be two triangles essentially different, each of which will answer the conditions of the problem; if it touch the semicircle, there will be only one triangle, and it will be the greatest possible with the base of the given magnitude, and the sides in the given ratio; and hence we have the means of solving the problem in which it is required *to construct a triangle on a given base, having its sides in a given ratio, and its area a maximum.*

Method of Computation. Join the center II with E, and draw the perpendicular EK. Then, let $m : n :: AC : CB$, and consequently $m : n :: AD : DB$; then, from these, by composition and division, we get $m+n : n :: AB : CB$, and $m-n : n :: AB : DB$; whence DC and its half, the radius of the circle, become known. EK also is found by dividing double of the area by AB. Then, in the triangle EKII, KH can be found, and thence AK and KB; and, by means of them and EK, the sides EA, EB may be computed. If E' be taken as the vertex, the method of computation is almost the same, and is equally easy.

Prob. XXIV.—Prob.— *Given the base of a triangle, the vertical angle, and the rectangle of the sides ; to construct it.*

On the given base describe a segment containing an angle equal to the given angle; to the diameter of the circle of which this segment is a part, and to the lines containing the given rectangle, find a fourth proportional; this proportional (Ex. 3) is the perpendicular of the triangle ; and the rest of the solution is effected by means of the fifteenth proposition of these Exercises.

Prop. XXV.—Prob.—*To divide a given triangle into two parts in a given ratio, by a straight line parallel to one of the sides.*

Let ABC be a given triangle; it is required to divide it into two parts in the ratio of the two straight lines, m, n, by a straight line parallel to the side BC.

Divide (V. 3, scho.) AB in G, so that $BG : GA :: m : n$, and between AB, AG find (V. 11) the mean proportional AII·

draw HK parallel to BC; ABC is divided by HK in the manner required.

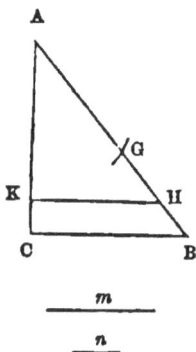

For (V. 14, scho.) since the three straight lines AB, AH, AG are proportionals, AB is to AG as the triangle ABC to AHK; whence, by division, BG is to AG, or (const.) *m* is to *n*, as the quadrilateral BCKH to the triangle AHK.

In practice, the construction is easily and elegantly effected, when the triangle is to be divided either into two or more parts proportional to given lines, by dividing AB into parts proportional to those lines, and through the points of section drawing perpendiculars to AB, cutting the arc of a semicircle described on AB as diameter; then by taking lines on AB, terminated at A, and severally equal to the chords drawn from A to the points of section of the arc, the points on AB will be obtained through which the parallels to BC are to be drawn. The reason is evident from the foregoing proof in connection with the corollary to the eighth proposition of the fifth book.

Cor. Hence a given triangle can be divided into two parts in a given ratio, by a straight line parallel to a given straight line.

Also, a triangle can be divided into two parts in a given ratio, by a straight line drawn through a given point in one of the sides; and a given quadrilateral can be divided into two parts in a given ratio, by a straight line parallel to one of its sides.

PROP. XXVI.—PROB.—*From a given point in one of the sides of a given triangle, to draw two straight lines trisecting the triangle.*

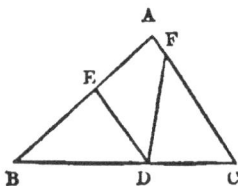

Let ABC be the given triangle, and D the given point; then, if BD be not less than one third of BC, to BD, to a third part of BC, and to the perpendicular from A to BC, find a fourth proportional, and at a distance equal to it draw a parallel to BC, cutting BA in E, and join ED ; the triangle BED is evidently (V. 10, cor., and I. 15, cor. 4) a third part of ABC ; and CDF will be constructed in a similar manner, if CD be not less than a third of BC.

If DC, one of the segments, be less than a third of BC, the triangle BDE is constructed as before, but the rest of the preceding solution fails, as the second parallel would fall above the triangle. In this case, cut off BA between E and A, a part equal to BE, and call it EG ; then, if DG be joined, the triangle ABC is trisected by DE, DG.

Scho. It is easy to see that this method may be readily extended to the division of a triangle into more equal parts than three, or into parts proportional to given magnitudes, by straight lines drawn from a given point in one of the sides.

PROP. XXVII.—THEOR.—*If the sides of a right-angled triangle be continual proportionals, the hypothenuse is divided in extreme and mean ratio by the perpendicular to it from the right angle ; and the greater segment is equal to the less or remote side of the triangle.*

Let ABC be a triangle right-angled at A, and let AD be perpendicular to BC ; then if CB : BA :: BA : AC, BC is divided in extreme and mean ratio in D, and BD is equal to AC.

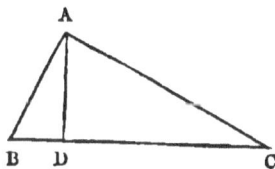

For (hyp.) CB : BA :: BA : AC, and (V. 8, cor.) CB : BA :: BA : BD ; therefore BA : AC :: BA : BD, and AC is equal to BD. Again (V. 8, cor.), BC : CA :: CA : CD, or BC : BD :: BD : DC, and therefore (V. def. 3) BC is divided in extreme and mean ratio in D.

Scho. Conversely, if BC : BD :: BD : CD, and if BAC be a right angle, and DA perpendicular to BC ; CB : BA :: BA : AC, and BD is equal to AC. For (hyp.) CB : BD :: BD : CD, and (V. 8, cor.) CB : CA :: CA : CD ; wherefore BD² is equal to CA², each (V. 9, cor.) being equal to the rectangle BC.CD, and therefore BD is equal to CA. Again (V. 8), CB : BA :: BA : BD or AC.

Prop. XXVIII.—Prob.—*Given the angles and diagonals of a parallelogram; to construct it.*

On one of the diagonals describe a segment of a circle containing an angle equal to the given angle at either extremity of the other; from the middle point of this diagonal as center, with half the other diagonal as radius, describe an arc cutting the arc of the segment; through the extremities of the first diagonal draw four straight lines, two to the intersection of the arcs, and two parallel to these; the parallelogram thus formed is easily proved to be the one required.

Prop. XXIX.—Prob.—*Given the vertical angle of a triangle, and the radii of the circles inscribed in the parts into which the triangle is divided by the perpendicular; to construct the triangle.*

Take any straight line ABC, and through any point B draw the perpendicular BD; make BA, BC equal to the given radii, and let E, F be the angular points, remote from B, of squares described on AB, BC; join EF, and on it describe the segment EDF, containing an angle equal to half the given vertical angle; let the perpendicular cut the arc EDF in D, and join DE, DF; draw DG, DH making the angles EDG, FDH respectively equal to EDB, FDB; DGH is the required triangle.

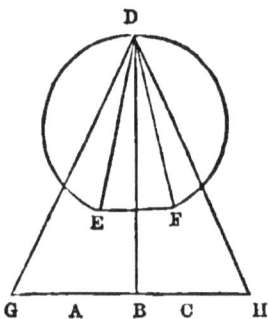

For (L 14) perpendiculars drawn from E to DB, DG are equal, and each of them is equal (const.) to the perpendicular from E to GB. Each of them therefore is equal to the given radius AB and a circle described from E at the distance of one of these is inscribed in the triangle DGB. In the same manner it would be shown, that a circle described from F as center, with the other given radius, would be inscribed in DBH. Hence, since the angle GDH is double of EDF, GDH is equal to the given vertical angle, and the triangle GDH answers the conditions of the question.

Scho. The preceding solution is strictly in accordance with the enunciation, taken in its limited sense. There will be interesting variations, however, if we regard the given circles, not merely as *inscribed*, but as those which touch all the sides of each of the right-angled triangles, either internally or externally. These variations will be obtained by giving the squares on the radii every possible variety of position in the four right angles formed by the intersection of AC, DB; and the solution will obtain complete generality by taking into consideration both the points in which BD cuts the circle of which EF is a chord.

PROP. XXX.—THEOR.—*The area of a triangle* ABC *is equal to half the continued product of two of its sides,* AB, BC, *and the sine of their contained angle* B, *to the radius* I.

Draw the perpendicular AD. Then (TRIG. 1, cor.) AD = AB × sin B. Multiply by BC, and take half the product; and (I. 23, cor. 6) we have the area equal to $\frac{1}{2}$ AB × BC × sin B.

Cor. Hence (I. 15, cor. 1) the area of a parallelogram is equal to the continual product of two contiguous sides, and the sine of the contained angle.

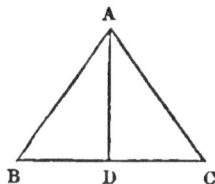

Scho. From this proposition, and from the third and sixth corollaries to the twenty-fifth proposition of the third book, we can derive neat algebraic expressions for the radii of the four circles, each touching the three sides of a triangle. Thus, by dividing the expression for the area by *s*, we find, according to the third corollary, that the radius of the inscribed circle is equal to $\sqrt{\dfrac{(s-a)\ (s-b)\ (s-c)}{s}}$. In like manner, by dividing the expression for the area successively by $s-a$, $s-b$, $s-c$, we find, according to the sixth corollary, that the radii of the circles touching a, b, c, externally, are respectively,

$$\sqrt{\frac{s\ (s-b)\ (s-c)}{s-a}},\ \sqrt{\frac{s\ (s-a)\ (s-c)}{s-b}},\ \text{and}\ \sqrt{\frac{s\ (s-a)\ (s-b)}{s-c}}.$$

By taking the continual product of these four expressions, and contracting the result, we get $s\ (s-a)\ (s-b)\ (s-c)$, which is equal to the square of the area; and hence, by expressing this inwards, we have the following remarkable theorem: *The continual product of the radii of the four circles, each of which touches the three sides of a triangle, or their prolongations, is equal to the second power of the area,* which is proven by the next proposition.

PROP. XXXI.—THEOR.—*Let* a, b, c *be the sides of a triangle, and* s *half their sum ; the area is equal to the square root of the continual product of* s, s—a, s—b, *and* s—c.

It was proved in the second corollary to the ninth proposition, Plane Trigonometry, that the sine of twice any angle is twice the product of the sine and cosine of the angle. Hence, by multiplying together the values of $\sin \frac{1}{2} A$ and $\cos \frac{1}{2} A$, given in the corollaries to the sixth and seventh propositions, Plane Trigonometry, and doubling the result, we get $\sin A = \dfrac{2 \sqrt{[s\,(s-a)\,(s-b)\,(s-c)]}}{bc}$. Now, by the preceding proposition, the area of a triangle is found by multiplying the sine of one of its angles by the sides containing it, and taking half of the product; multiplying, therefore, the value now found for $\sin A$, by bc, and taking half the product, we find the area to be $\sqrt{[s\,(s-a)\,(s-b)\,(s-c)]}$. This proposition is much used in surveying coasts and harbors.

PROP. XXXII.—PROB.—*A semicircle* ACB *being given, and other semicircles being described as in the diagram ; it is required to find the sum of the areas of all those inscribed semicircles.*

Circles (V. 14), and consequently semicircles, are as the

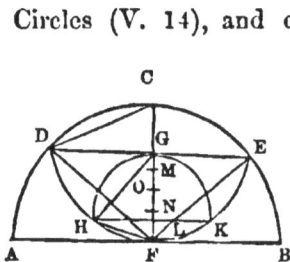

squares of their diameters or of their radii. Now the square of GD is half the square of DF or CF, and therefore the semicircle DFE is half of ACB. For the same reason HGK is half of DFE; and universally, each semicircle is half of the one in which it is inscribed. Hence the entire amount will be the sum of the infinite series $\frac{1}{2} ACB + \frac{1}{4} ACB + \frac{1}{8} ACB + \frac{1}{16} ACB + $ etc. ; and therefore (IV. 19) $\frac{1}{2}$ ACB $-\frac{1}{4}$ ACB : $\frac{1}{2}$ ACB :: $\frac{1}{2}$ ACB : ACB, the required sum; and it thus appears that the sum of all the inscribed semicircles is equivalent to the given semicircle.

PROP. XXXIII.—THEOR.—*In any triangle, the center of the circumscribed circle, the point in which the three perpendiculars intersect one another, and the point of intersection of the straight lines drawn from the angles to bisect the opposite sides, lie all in the same straight line.*

Let ABC be a triangle, and let the two perpendiculars AD, CE intersect in F; bisect AB, BC in H, G, and draw AG, CH intersecting in K; draw also GI, HI perpendicular to BC, BA, and intersecting in I. Then (Ex. 7) F is the intersection of the three perpendiculars, K (III. 1, cor. 2) the intersection of the three lines drawn from the angles to bisect the opposite sides, and (III. 25, cor. 2) I is the center of the circumscribed circle. Join FK, KI; FKI is a straight line.

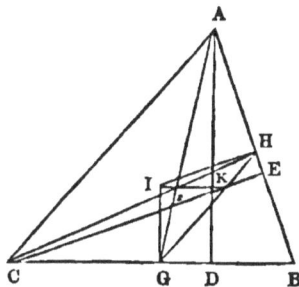

Join HG; it is (V. 2 and 3) parallel to AC, and is half of it. Also the triangles ACF, GHI are (I. 16, cor. 3) equiangular, and therefore GI is half of AF. So likewise (III. 1, scho.) is KG of KA. Hence the two triangles AKF, GKI have the alternate angles KAF, KGI equal, and the sides about them proportional; therefore (V. 6) the angles AKF, GKI are equal, and (I. 10, cor.) since AKG is a straight line, FKI is also a straight line.

Scho. It is plain (V. 3) that FK is double of KI. We have also seen that AF is double of GI. Hence it appears, that the distance between any of the angles and the point of intersection of the three perpendiculars is double of the perpendicular drawn from the center of the circumscribed circle to the side opposite to that angle.

PROP. XXXIV.—THEOR.—*Straight lines drawn from the angles of a triangle to the points in which the opposite sides touch the inscribed circle, all pass through the same point.*

Let ABC be a triangle, and D, E the points in which the sides AB, AC touch the inscribed circle; draw BFE, CFD; draw also AFG cutting BC in G; G is the point in which BC touches the inscribed circle.

If possible, let another point K be the point of contact, and draw DII, DI parallel to BC, CA. Then in the similar triangles FDI, FCE, FD : DI :: FC : CE, or CK; and in the similar triangles FDII, FGC, DII : DF :: GC : FC; from which and from the preceding analogy we get, *ex æquo*, DH : DI :: CG : CK. Again, BD : DI :: BA : AE or AD :: BG : DII. Hence, alternately, and by inversion, BG : BD :: DII : DI : whence (IV. 7) BG : BD or BK :: CG : CK, or alternately, BG : CG :: BK : CK; and by composition, BC : CG :: BC : CK; and therefore CG, CK are equal; that is, G and K coincide, and AFG passes through the point in which BC touches the circle.

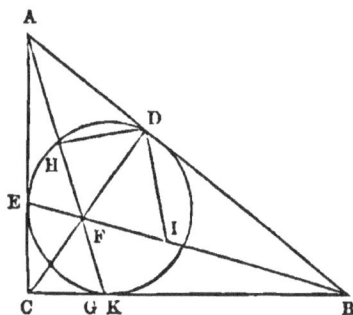

PROP. XXXV.—THEOR.—*In a triangle, the sum of the perpendiculars drawn from the center of the circumscribed circle to the three sides is equal to the sum of the radii of the inscribed and circumscribed circles.*

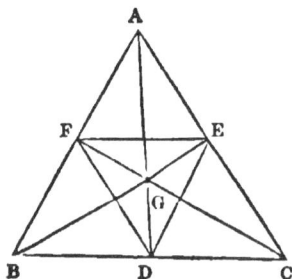

Let ABC be a triangle, having its sides bisected in D, E, F, by perpendiculars meeting in G, the center of the circumscribed circle; the sum of GD, GE, GF is equal to the sum of the radii of the inscribed and circumscribed circles.

Join GA, GB, GC, and DE, EF, FD. Then, putting a, b, c to denote the sides opposite to the angles A, B, C, we have (V. 2 and 3) FE $=\frac{1}{2}a$, FD $=\frac{1}{2}b$, and DE $=\frac{1}{2}c$; and (III. 11) since AEG, AFG are right angles, a circle may be described about the quadrilateral AEGF. For a like reason circles may be described about BDGF and CDGE. Hence (Ex. 4) FE. AG$=$AF.GE$+$AE.FG ; or, by doubling, a.AG$=c$.GE$+b$.FG. In the same manner, it would be shown, since AG, BG, CG are equal, that b.AG$=c$.GD$+a$.FG, and c.AG$=a$.GE$+b$.GD. Hence, by addition, $(a+b+c)$AG$=(a+c)$GE$+(a+b)$GF$+(b+c)$GD. Now b.GE is evidently equal to twice the triangle AGC, c.GF equal to twice AGB, and a.GD equal to twice BGC; also, denoting the radius of the inscribed circle by r, we have (III. 25, cor. 3), $r(a+b+c)$ equal to twice the area of the triangle ABC, and consequently $r(a+b+c)=b$.GE$+c$.GF $+a$.GD. Hence, by addition, $(a+b+c)$AG$+r(a+b+c)=$ $(a+b+c)$GE$+(a+b+c)$GF$+(a+b+c)$GD; and consequently AG$+r=$GE$+$GF$+$GD.

Cor. Since, by the scholium to proposition thirty-third of these Exercises, the parts of the three perpendiculars of the triangle, between their common intersection and the three angles, are respectively double of GD, GE, GF, the sum of those parts of the perpendiculars is equal to the sum of the diameters of the inscribed and circumscribed circles.

PROP. XXXVI.—THEOR.—*If on the three sides of any triangle equilateral triangles be described, either all externally, or all internally, straight lines joining the centers of the circles inscribed in those three triangles form an equilateral triangle.*

On the three sides of any triangle ABC, let the equilateral triangles ABD, BCF, CAE be described externally, and find G, H, K, the centers of the circles described in those triangles ; draw GH, HK, KG ; GHK is an equilateral triangle.

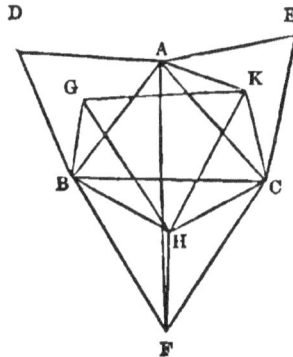

Join GA, GB, HB, HC, HF, KC, AF. Then the angle FBC is two thirds of a right angle, and the angles GAB, GBA, FBH, BFH each one third. The triangles FBH, ABG are therefore similar, and (V. 3) FB : BH :: BA : BG ; whence, alternately, FB : BA ::

HB : BG; that is, in the triangles FBA, HBG the sides about
the angles FBA, HBG are proportional; and these angles are
equal, each of them being equal to the sum of the angle ABC
and two thirds of a right angle. Hence (V. 6) these triangles
are equiangular; and therefore (V. 3) FB or BC : FA :: BH
: HG; and it would be shown in the same manner, by means
of the triangles ACF, KCH, that FC or BC : FA :: CH :
HK; therefore (IV. 7) BH : HG :: CH : HK. But BH is
equal to CH, and therefore (IV. 6) HG is equal to HK; and it
would be demonstrated in a similar manner, that HG, HK are
each equal to GK.

If the equilateral triangles were described on the other sides
of the lines AB, BC, CA, the angles ABF, GBH would be the
difference between ABC and two thirds of a right angle; but
the rest of the proof is the same.

Scho. If ABC exceed a right angle and a third, the sum of it
and two thirds of a right angle is greater than two right angles.
In that case, the angles ABF, GBH, understood in the ordinary
sense, are each the difference between that sum and four right
angles. If ABC be a right angle and a third, the sum becomes
two right angles, and FB, BA are in the same straight line, as
are also HB, BG. In this case it is proved as before, that FB :
BA :: HB : BG; and then (IV. 11) FB or BC : FA :: HB :
HG. The rest of the proof would proceed as above.

It may be remarked that if an equilateral triangle be described on a straight line, and if on the two parts into which it
is divided at any point, other equilateral triangles be described,
lying in the opposite direction, the lines joining the centers of
the three equilateral triangles will also form an equilateral triangle. The connection of this and the proposition will be perceived by supposing two angles of the triangle continually to
diminish, till they vanish, as the triangle may thus be conceived
to become a straight line.

TO BE PROVEN.

1. The least straight line that can be drawn to another
straight line from a point without it, is the perpendicular to it;
of others, that which is nearer to the perpendicular is less than

one more remote; and only two equal straight lines can be drawn, one on each side of the perpendicular.

2. Of the triangles formed by drawing straight lines from a point within a parallelogram to the several angles, each pair that have opposite sides of the parallelogram as bases, are half of it.

3. If, in proceeding round an equilateral triangle, a square, or any regular polygon, in the same direction, points be taken on the sides, or the sides produced, at equal distances from the several angles, a similar rectilineal figure will be formed by joining each point of section with those on each side of it.

4. If the three sides of one triangle be perpendicular to the three sides of another, each to each, the triangles are equiangular.

5. A trapezoid, that is, a trapezium having two of its sides parallel, is equivalent to a triangle which has its base equal to the sum of the parallel sides, and its altitude equal to their perpendicular distance.

6. Given the segments into which the line bisecting the vertical angle divides the base, and the difference of the angles at the base; to construct the triangle, and compute the sides.

7. Within or without a triangle, to draw a straight line parallel to the base, such that it may be equivalent to the parts of the other sides, or of their continuations, between it and the base.

8. Given the perpendicular of a triangle, the difference of the segments into which it divides the base, and the difference of the angles at the base; to construct the triangle.

9. In the figure for the first corollary to the twenty-fourth proposition of the first book, prove that CD is perpendicular to AH, or a line from C to F is perpendicular to CD.

10. The angle made by two chords of a circle, or by their continuations, is equal to an angle at the circumference standing on an arc equivalent to the sum of the arcs intercepted between the chords, if the point of intersection be within the circle, or to their difference, if it be without; (2) the angle made by a tangent and a line cutting the circle, is equal to an angle at the circumference on an arc equivalent to the difference of the arcs intercepted between the point of contact and

the other line; and (3) the angle made by two tangents is equal to an angle at the circumference standing on an arc equivalent to the difference of those into which the circumference is divided at the points of contact.

Cor. If a tangent be parallel to a chord, the arcs between the point of contact and the extremities of the chord are equal.

11. Given the sum of the perimeter and diagonal of a square; to construct it.

12. On a given straight line describe a square, and on the side opposite to the given line describe equilateral triangles lying in opposite directions; circles described through the extremities of the given line, and through the vertices of these triangles, are equal.

13. To inscribe an equilateral triangle in a given square.

14. Given the angles and the two opposite sides of a trapezium; to construct it.

15. In a given circle to place two chords of given lengths, and inclined at a given angle.

16. In the figure for the twenty-fifth proposition of the third book, if AM, BM be joined, the angle AMB is half of ACB.

17. If, in proceeding in the same direction round any triangle, as in 3, points be taken at distances from the several angles, each equal to a third of the side, the triangle formed by joining the points of section is one third of the entire triangle.

18. To describe a square having two of its angular points on the circumference of a given circle, and the other two on two given straight lines drawn through the center. Show that there may be eight such squares.

19. Given the vertical angle of a triangle, and the segments into which the base is divided at the point of contact of the inscribed circle; to describe the triangle, and compute the sides.

20. If any three angles of an equilateral pentagon be equal, all its angles are equal.

21. Given two sides of a triangle, and the difference of the segments into which the third side is divided by the perpendicular from the opposite angle; to construct the triangle.

22. Given the vertical angle of a triangle, the line bisecting it, and the perpendicular; to construct the triangle.

23. From a given center to describe a circle, from which a

straight line, given in position, will cut off a segment containing an angle equal to a given angle.

24. In a given triangle to inscribe a semicircle having its center in one of the sides.

25. Through three given points to draw three parallels, two of which may be equally distant from the one between them.

26. Given an angle of a triangle, and the radii of the circles touching the sides of the triangles into which the straight line bisecting the given angle divides the triangle; to construct it.

27. In a rhombus to inscribe a square.

28. Given the lengths of the two parallel sides of a trapezoid, and the lengths of the other sides; to construct it.

29. Given one of the angles at the base, and the segments into which the base is divided at the point of contact of the inscribed circle; to describe the triangle.

30. To draw a tangent to a given circle, such that the part of it intercepted between the continuations of two given diameters may be equal to a given straight line.

31. To draw a tangent to a given circle, such that the part of it between two given tangents to the circle may be equal to a given straight line.

32. Given the vertical angle of a triangle, the difference of the sides, and the difference of the segments into which the line bisecting the vertical angle divides the base; to construct the triangle.

33. Given any three of the circles mentioned in the fifth corollary to the twenty-fifth proposition of the third book; to describe the triangle.

34. A straight line and a point being given in position, it is required to draw through the point two straight lines inclined at a given angle, and inclosing with the given line a space of given magnitude.

35. From two given straight lines to cut off equal parts, each of which will be a mean proportional between the remainders.

36. A square is to a regular octagon described on one of its sides, as 1 to 2 $(1 + \sqrt{2})$.

37. In a given triangle to inscribe a parallelogram of a given area.

38. In a given circle to inscribe a parallelogram of a given area.

39. Through a given point between the lines forming a given angle, to draw a straight line cutting off the least possible triangle.

40. To divide a given straight line into two parts, such that the square of one of them may be double of the square of the other, or may be in any given ratio to it.

41. To produce a given straight line, so that the square of the whole line thus produced may be double of the square of the part added, or in any given ratio to it.

42. Given the area of a right-angled triangle, and the sum of the legs ; to construct it.

43. Given the area and the difference of the legs of a right-angled triangle; to construct it.

44. Given one leg of a right-angled triangle, and the remote segment of the hypothenuse, made by a perpendicular from the right angle; to construct the triangle.

45. Given the base of a triangle, the vertical angle, and the side of the inscribed square standing on the base; to describe the triangle.

46. Given the base of a triangle, and the radii of the inscribed and circumscribed circles ; to construct the triangle.

47. To draw a chord in a circle which will be equal to one of the segments of the diameter that bisects it.

48. In a given circle to draw a chord which will be equal to the difference of the parts into which it divides the diameter that bisects it.

49. If two sides of a regular octagon, between which two others lie, be produced to meet, each of the produced parts is equivalent to a side of the octagon, together with the diagonal of a square, described on the side.

50. The perimeter of a triangle is to the base as the perpendicular to the radius of the inscribed circle.

51. From a given point without a given circle to draw a straight line cutting the circle, so that the external and internal parts may be in a given ratio.

52. Each of the complements of the parallelograms, about

the diagonal of a parallelogram, is a mean proportional between those parallelograms.

53. Given the ratio of two straight lines, and the difference of their squares; to find them.

54. The square of the perimeter of a right-angled triangle is equivalent to twice the rectangle under the sum of the hypothenuse and one leg, and the sum of the hypothenuse and the other.

55. The quadrilateral formed by straight lines bisecting each pair of adjacent sides of a quadrilateral is a parallelogram, which is half of the quadrilateral; and straight lines, joining the points in which the sides of that parallelogram are cut by the diagonals of the primitive figure, form a quadrilateral similar to that figure and equivalent to a fourth of it.

56. From three given points as centers, and not in the same straight line, to describe three circles each touching the other two. Show that this admits of four solutions.

57. To add a parallelogram to a rhombus, such that the whole figure may be a parallelogram similar to the one added.

58. A straight line being given in position, and a circle in magnitude and position, it is required to describe two equal circles touching one another, and each touching the straight line and the circle.

59. The squares of the diagonals of a quadrilateral are together double of the squares of the straight lines joining the points of bisection of the opposite sides.

60. In a given rhombus to inscribe a rectangle having its sides in a given ratio.

61. If, through the vertex and the extremities of the base of a triangle, two circles be described intersecting one another in the base or its continuation, their diameters are proportional to the sides of the triangle.

62. To draw a straight line cutting two given concentric circles, so that the parts of it within them may be in a given ratio.

63. From a given point, within a given circle, or without it, to draw two straight lines to the circumference, perpendicular to one another, and in a given ratio. When will this be impossible?

17

64. If a straight line be divided in extreme and mean ratio, the squares of the whole and the less part are together equivalent to three times the square of the greater.

65. If a straight line be cut in extreme and mean ratio, and be also bisected, the square of the intermediate part, and three times the square of half the line are equivalent to twice the square of the greater part.

66. If the hypothenuse of a right-angled triangle be given, the side of the greatest inscribed square, standing on the hypothenuse, is one third of the hypothenuse.

67. To divide a given semicircle into two parts by a perpendicular to the diameter, so that the radii of the circles inscribed in them may be in a given ratio.

68. To draw a straight line parallel to the base of a triangle, making a segment of one side equal to the remote segment of the other.

69. In the figure for the tenth proposition of the second book, the square of the diameter of the circle, passing through the points F, II, D, is six times the square of the straight line joining FD.

70. Given the base, the area, and the sum of the squares of the sides of a triangle; to construct it.

71. If, from the extremities of the hypothenuse of a right-angled triangle as centers, arcs be described passing through the right angle, the hypothenuse is divided into three segments, such that the square of the middle one is equivalent to twice the rectangle of the others.

72. On a given hypothenuse to describe a right-angled triangle, such that the difference between one leg and the adjacent segment of the hypothenuse made by a perpendicular from the right angle, may be a maximum.

73. On a given hypothenuse to construct a right-angled triangle, such that one segment of the hypothenuse made by the perpendicular from the right angle, may be equivalent to the sum of the perpendicular and the other segment.

74. The square of DII (see figure for III. 24) is equivalent to the rectangle CF.BG ; and if the circles touch one another externally, DII is a mean proportional between their diameters. Also the square of DII is equivalent to the rectangle CG.BF.

75. On a given straight line to describe an isosceles triangle, having the vertical angle treble of each of the angles at the base.

76. If in the diameter of a circle and its continuation two points be taken on the opposite sides of the center, such that the rectangle under their distances from the center may be equivalent to the square of the radius, any circle whatever described through these points bisects the circumference of the other circle.

77. To find a point from which, if straight lines be drawn to three given points, they will be proportional to three given straight lines.

78. Given the segments into which the base of a triangle is divided by two straight lines trisecting the vertical angle; to construct it.

79. If one diagonal of a quadrilateral inscribed in a circle be bisected by the other, the square of the latter is equivalent to half the sum of the squares of the four sides.

80. To divide a straight line into two parts, such that the squares of the whole and one of the parts may be double of the square of the other part.

81. Given the segments into which the base of a triangle is divided by the straight line bisecting the vertical angle, to construct the triangle so that its angle adjacent to the greater segment may be either of a given magnitude, or a maximum, and in each case to compute the remaining sides and angles.

82. To draw a straight line bisecting a given parallelogram, so that if it be produced to meet the sides produced, the external triangles will have a given ratio to the parallelogram.

83. Through a given point to draw a straight line, which, if continued, would pass through the point of intersection of two given inclined straight lines, without producing those lines to meet.

84. If, from any point in the circumference of the circle described about an equilateral triangle, chords be drawn to its three angles, the sum of their squares is equivalent to six times the square of the radius of the same circle, or to twice the square of a side of the triangle.

85. Given the difference of the angles at the base of a trian-

gle, the difference of the segments into which the base is divided by the perpendicular, and the ratio of the sides; to construct the triangle.

86. Given the base and vertical angle of a triangle, to construct it so that the line bisecting the vertical angle may be a mean proportional between the segments into which it divides the base.

87. Given two sides of a triangle, and the ratio of the base and the line bisecting the vertical angle; to construct the triangle.

88. If the vertical angle of a triangle be double of one of the angles at the base, the rectangle under the sides is equivalent to the rectangle under the base, and the line bisecting the vertical angle.

89. If a straight line be divided into parts, which, taken in succession, are continual proportionals, and if circles be described on the several parts as diameters, a straight line which touches two of the circles on the same side of the straight line joining their centers, will touch all the others.

90. Given the segments into which the base is divided by the straight line bisecting the vertical angle, and the angles which that straight line makes with the base; to construct the triangle.

91. To divide a given circle into two segments, such that the squares inscribed in them may be in a given ratio.

92. Through a given point in the base of a given isosceles triangle, or its continuation, to draw a straight line such that the lines intercepted on the equal sides, or their continuations between that line and the extremities of the base, may have one of the equal sides as a mean proportional between them.

93. Through a point in the circumference of a given circle, to draw two chords, such that their rectangle may be equivalent to a given space, and the chord joining their other extremities equal to a given straight line.

94. In any triangle the radius of the circumscribed circle is to the radius of the circle which is the locus of the vertex, when the base and the ratio of the sides are given, as the difference of the squares of those sides is to four times the area.

95. The difference of the sides of a triangle is a mean pro-

portional between the difference of the segments into which the base is divided by the perpendicular, and the difference of those into which it is divided by the line bisecting the vertical angle.

96. Let the angles of a parallelogram which has unequal sides be bisected by straight lines cutting the diagonals, and let the points of intersection be joined; the figure thus formed is a parallelogram, which has to the proposed parallelogram the duplicate ratio of that which the difference of the unequal sides of the latter has to their sum.

97. A circle and a point being given, it is required to describe a triangle similar to a given one, having its vertex at the given point, and its base a chord of the given circle.

98. Given the three points in which the sides of a triangle are cut by the perpendiculars from the opposite angles; to construct the triangle.

99. Given the angles of a triangle, and the lengths of three straight lines drawn from the angular points to meet in another point; to construct the triangle.

100. Given the base of a triangle, and the ratio of its sides; to construct it, so that the distance of its vertex from a given point may be a maximum or minimum.

101. To divide a circle into two segments, such that the sum of the squares inscribed in them may be equivalent to a given space.

102. Through a given point, with a given radius, to describe a circle bisecting the circumference of a given circle.

103. With a given radius to describe a circle bisecting the circumferences of two given circles.

104. In a right-angled triangle, the rectangle under the radius of the inscribed circle, and the radius of the circle touching the hypothenuse and the legs produced, is equivalent to the area. So, likewise, is the rectangle under the circles touching the legs externally, and the continuations of the other sides.

105. If three straight lines be continual proportionals, the sum of the extremes, their difference, and double the mean will be the hypothenuse and legs of a right-angled triangle.

106. From two given centers, to describe circles having their

radii in a given ratio, and the part of their common tangent, between the points of contact, equal to a given straight line.

107. A straight line and two points equally distant from it, on the same side, being given in position, it is required to draw through the points two straight lines forming with the given line the least isosceles triangle possible, on the side on which the points are.

108. To describe a circle touching a diameter of a given circle in a given point, and having its circumference bisected by that of the given one.

109. If an angle of a triangle be 60°, the square of the opposite side is less than the squares of the other two by their rectangle; but if an angle be 120°, the square of the opposite side is greater than the squares of the others by their rectangle.

110. In the figure on page 251, prove that the three straight lines joining AF, BE, CD are all equal.

111. The chord of 120° is equal to the tangent of 60°.

112. The sines of the parts into which the vertical angle of a triangle is divided by the straight line bisecting the base, are reciprocally proportional to the adjacent sides. Show from this how a given angle may be divided into two parts, having their sines in a given ratio.

113. The diameter of the circle described about any triangle is equivalent to the product of any side and the cosecant of the opposite angle.

114. In any triangle ABC, the radius of the inscribed circle is equivalent to $a.\dfrac{\sin \frac{1}{2} B \sin \frac{1}{2} C}{\cos \frac{1}{2} A}$, to $b.\dfrac{\sin \frac{1}{2} A \sin \frac{1}{2} C}{\cos \frac{1}{2} B}$, or to $c.\dfrac{\sin \frac{1}{2} A \sin \frac{1}{2} B}{\cos \frac{1}{2} C}$; or, finally, to the cube root of

$abc \sin \frac{1}{2} A \sin \frac{1}{2} B \sin \frac{1}{2} C \tan \frac{1}{2} A \tan \frac{1}{2} B \tan \frac{1}{2} C$.

115. Given the sum of the tangents, and the ratio of the secants, of two angles to a given radius; to determine the angles geometrically and by computation.

116. Find an angle, such that its tangent is to the tangent of its double, in a given ratio; suppose that of 2 to 5.

117. If a spherical triangle be right-angled at C, and any two of the other parts be made successively 108° 42′ and 87° 33′ 19″, what will be the values of the remaining parts?

118. Let any three parts of an oblique-angled spherical triangle be successively 87° 45′, 96° 57′ 48″, and 106° 53′ 13″; what will be the values of the other parts?

119. When the three sides of a spherical triangle be respectively, 34° 39′ 44″, 78° 27′ 49″, and 134° 15′ 23″, what will be the surface of the triangle when the radius of the sphere is 16? What will be the base of the triangular pyramid subtended by those sides, and what will be the surface and solidity of the spherical pyramid with its apex at the center of the sphere?

120. If the foregoing values be the magnitudes of the angles of a spherical triangle when the radius of the sphere is 16, what will be the base of the triangular pyramid subtended by the sides of the spherical triangle? What will be the surface of the spherical triangle, and the surface and solidity of the spherical pyramid with its apex at the center of the sphere?

121. If the earth be regarded a sphere with 7973.8798+ miles for its diameter, and a spherical pentagon be measured on its surface having 341.78 miles, 309.25 miles, 278.64 miles, 173.97 miles, and 97 miles for its sides; and the angles contained by those sides be respectively 74° 34′ 19″, 107° 09′ 51″, 41° 0′ 11″, 85° 17′ 09″, and 76° 41′ 35″, what will be the surface of the spherical pentagon? and what will be the solidity of the pyramid having the spherical pentagon for its base and its apex at the center of the earth?

ADDENDUM,

Illustrating the Third Proof for the Second Corollary to the Seventeenth Proposition of the Sixth Book, Elements of Euclid and Legendre.

Circles are to one another as the squares described upon their diameters (V. 14), consequently squares are to one another as the circles described (V. 14, cor. 2) upon their sides; that is, there is a ratio of equality between the circle and squares which have the same straight line for their respective diameter, side, and diagonal· therefore the circle has the same arithmetical proportion to the inscribed square having the diameter for its diagonal, as the circumscribed square having the same diameter for its side has to the circle. If 10 be the diameter of the circle, and χ be the area of the circle, 100 will be the area of the circumscribed square, and 50 will be the area of the inscribed square—and we have the arithmetical proportion,

$$100, \chi, 50;$$

and the geometrical proportion,

$$100 : \chi :: 50 : \frac{\chi}{2}.$$

From the first proportion, we derive

$$100 - \chi = \chi - 50.$$

From the second proportion, we have

$$100 - \chi = 2\left(50 - \frac{\chi}{2}\right).$$

$$\therefore \chi - 50 = 2\left(50 - \tfrac{1}{2}\chi\right) = 100 - \chi;$$

$$\text{or, } 2\chi = 150, \text{ hence } \chi = 75.$$

The arithmetical proportion gives,

$$2\chi = 150 \text{ or } \chi = 75$$

Substituting this value for χ in the above proportions, we get

$$100 : 75 :: 50 : 37\tfrac{1}{2};$$
$$\text{and } 100 - 75 = 75 - 50;$$
$$\text{and } 100 - 75 = 2\left(50 - 37\tfrac{1}{2}\right);$$
$$\text{and } 100, 75, 50;$$

$$\text{or, } 2\,(75) = 150, \text{ or } 75 = \frac{150}{2} = 75.$$

Hence, the circle is the *arithmetical mean* between the squares, circumscribed and inscribed about it; or three fourths of the circumscribed square; or three times square of the radius. (See Exercises, def. 7, scho. 2).

Thus we have the area of the circle expressed by a *finite* quantity instead of the *irrational* quantity (V. 25, scho. 1), giving the *approximate* area only of the circle.